UFO Encounters

How High Strange Events Transform Human Perception

"The most beautiful thing we can experience is the mysterious.
It is the source of all true art and science."
-Albert Einstein

Angelia Sheer, BS, MBA

CONTENTS

Acknowledgments   i

Foreword   v

Introduction   ix

1 Military Sightings   1

    No One Saw A Thing – UFO At Sea

2 Strange Entities   11

    The Captain and the Protector

    Late Night Car Visit

    Walking Through Walls

    Peeking Aliens I

    Peeking Aliens II

3 UFO Sightings: Lives Changed Forever   51

    UFO Flyby Witness "D"

    A Drive In The Park

    Journals

4 Missing Time   87

    On The Railroad Tracks

    Strange Voices

The Strange Case of Witness "J"

5   Field Investigation Sightings   131

    Transforming Lights

    Does Anyone Have A Camera?

6   Crossover Events: Hunters, Cryptids, Strange Lights   157

    I Don't Have A Flashlight

    The Deer That Wasn't

    Bright Lights and Missing Ammo

7   Noteworthy Paranormal Events   185

    Reading Minds: The Case For Telepathy

    Going Up, Please

    Strange Lights In The Bedroom

8   Perception "101"   205

9   Tying It All Together   211

    References   233

    About The Author   235

ACKNOWLEDGMENTS

I want to personally thank every witness that I have spoken with over the years. Their courage in the face of overwhelming odds to bring their truth forward is truly inspirational. I have been honored to be a part of their joys, fears, sorrows, triumphs, and, in many cases, radical transformations. I still marvel over the awe and wonder we have discovered together on our adventures and the fact that we have been given a small glimpse at the extraordinary creation we are all a part of. The artwork for the cover of the book was created by the great **Claudio Bergamin**, but its creation was inspired by many witness drawings and accounts. One piece of artwork done by experiencer, **Aaron Hill**, was so moving in its depth, I used it as my major influence for the cover of the book. Its startling nature and poignancy really conveyed the journey that so many of my brave witnesses embark on. Thank you, Aaron Hill, for your bravery in coming forward and for your beautiful artwork! Much love to my UFO family!

Next, I would like to thank **Mutual UFO Network (MUFON)**, its leadership, staff, and my many co-investigators for allowing me to be part of such a great family. I want to personally thank the following:

Jan Harzan, *Executive Director*

Thanks for being there night and day, always having a positive word of encouragement, and for sitting in the "hot seat" at times so we don't have to! Thanks, Jan, for all you do!

Steve Hudgeons, *Director of International Investigations*

Marketa Klimova, *Office Manager*

What would we ALL do without Marketa? You are appreciated everyday…thank you!

And a very special thank you to the following whose help, advice, guidance, and companionship are truly indispensable:

Barry Gaunt, *State Director, Kentucky*

Nancy DeSousa, *Assistant State Director, Kansas, and a member of the Experiencers Research Team (ERT)*

I also want to thank my family and all my friends for loving UFOgirl just as she is. I have so many stories of cards, letters, articles, videos, etc. that have been forwarded to me over the years by my clan, telling of strange sightings and events. I have one memory of my daughter that is just priceless. I was called out one evening regarding a sighting about an hour away. My regular research partner at the time was unable to go, so my daughter, **Shanna Simpson**, 16 at the time, volunteered to come along. We loaded all of the equipment, boots, extra clothes (it was freezing cold), snacks, and paperwork in the 4-wheel drive and set off. When we reached the site, we gathered up all of the equipment, suited up, and then set out to explore the sighting location. It was at this point my daughter stopped me. She walked over, took off her winter scarf, and meticulously tied one end to my belt loop and the other to her belt loop. I looked at her, perplexed, and she said, "They got me, they got you!" Thank you, my smart, beautiful girl!

Last but not least, a huge thank you to my partner in crime, **Dale Houston**. Dale grew up on the road with "Merle", still sings those old country songs, and travels all over the country on the #bigsilverbus. Every year he does charity events for St. Jude, manages his very busy tour bus business, rescues stranded musicians, and writes, records, and sings all over the country. Even in the midst of all of this, he still finds time to build my websites, design all of my promotional materials, take and edit my video, share many of my investigations via Facebook Live (he's the comic relief), and follow me around the country on speaking events and TV shows. Much love and gratitude!

- Angelia Sheer, UFOgirl, Tennessee

FOREWORD

When my story first became public, I would occasionally go on speaking engagements with Budd Hopkins. After his book, *Intruders, The Incredible Visitations at Copley Woods*, was published, I would join him on stage. Eventually, I accepted invitations to speak on my own. It wasn't easy for me, but I felt it was important. I am "Kathy Davis", the central figure in Budd's book, and now author of my own book, *Abducted, The Story of the Intruders Continues*. I felt I wanted to be as supportive of Budd as he had been of me and my family. And I wanted to pay forward some of that support to others who had experienced similar events in their lives. Many years have since passed, and I very rarely accept speaking engagements anymore. But when I was asked to speak in Eureka Springs, Arkansas, at the Ozark Mountain UFO Conference in 2017, I felt that compulsion to go that I can never resist, and I accepted immediately. I already knew that this was going to be a very special conference as it was also their 30th anniversary conference. (The Ozark Mountain UFO Conference is one of the longest-running UFO conferences in the world.)

I had previously met Angelia more than 20 years ago, at another UFO-related event, and we had, at that time, made a connection. But after the event, we lost touch with each other. I was very excited and happy to see her again in Eureka Springs. We picked up where we had left off all those years ago, as if we had just seen each other last week. When we reconnected, I immediately realized why I needed to be at this conference, at that time. In the years between, both of us had gotten more into the UFO research and investigation. Angelia became Chief Investigator and State Director for Tennessee MUFON and had started her own group, *Parasheer Research*, which not only studied the UFO phenomena but also investigated crossover events into other areas of inquiry.

Since my own experiences contained so many high strange events, I also began research into other forms of paranormal phenomena that seemed to inevitably cross over into the UFO mystery. Like me, Angelia had a strong feeling that many of these different kinds of paranormal events had some kind of connection, were often intertwined, and she wanted to know what that connection was. Oftentimes, at least in the past, many investigators that I have had contact with had limited themselves to just the "nuts and bolts" of UFO reports. They tended to dismiss reports of other paranormal activity when it crossed over into UFO reports. And they were even leerier of addressing the psychological, intellectual, and, even at times, spiritual aspect of these experiences. As you will read in Angelia's book, the UFO experience is often so much more than just "nuts and bolts". The effects it can have on the Experiencer can be profound and run the gamut from total denial and complete emotional shutdown to spiritual enlightenment and a feeling of some higher purpose that affects every aspect of their lives.

Thankfully Angelia has tuned in to this, and this is what makes her research unique and SO important! Her investigative technique is meticulous, and she leaves no stone unturned. Her vast knowledge on a wide range of subjects, including but not limited to psychology, hypnotherapy, cryptozoology, issues of spirituality, and her background in science, psychology, and medicine definitely gives her an edge. Nothing gets past her. Her enthusiasm for her research is contagious. Her compassion for her witnesses is obvious with every report she logs. And she is looking at things that other researchers shy away from. As she recalls each case outlined in this book, you, the reader, can see why her work is so vital to Ufology. She is on a mission to connect ALL the dots and find the truth. If anyone can do it, it is going to be Angelia. She is taking Ufology to the next level. I am so glad she finally had time to write this book. I believe it is going to be an important one for anyone with the desire to research UFOs and other paranormal activity. A must read for anyone interested in researching UFOs and the paranormal and anyone who has the desire to know the rest of the story. And I am certain it is only the first of hopefully many more books to come.

June 12, 2019

Debbie Jordan-Kauble, AKA Kathie Davis, From Budd Hopkins'
"Intruders, The Incredible Visitation at Copley Woods"

INTRODUCTION

"Only by confronting and yielding to the unknowable – by rigorously avoiding both the temptation to deny or explain away these phenomena or to try to find some conventional explanation for them – can we, as a species, evolve..."   - Kenneth Ring, Ph.D.

For all of my witnesses:

"Great spirits have always encountered violent opposition from mediocre minds."   - Albert Einstein

"250 billion+ stars in the Milky Way Galaxy, 100 billion estimated planets...are we alone? Doubtful..."   - Angelia Sheer

Over 35 years of boots on the ground investigations, more than 2000 witness interviews, and countless hours of research have led me to some intriguing theories about the UFO phenomena and the individuals experiencing their mystery. One of the most mind-bending challenges of our time has grudgingly yielded a glimpse of the complexity of our reality and of the integral part humanity plays in the understanding of these phenomena.

Over the years, UFO research has been mainly focused on the external phenomena at hand using the current scientific models. The sightings, the abductions, the radar echoes, physical traces, photos, videos, and search for that perfect piece of evidence were our holy grail. It didn't really seem to dawn on many researchers that our understanding of science is in no way complete and that there are other very valid methods of inquiry that may reveal to us a different interpretation of reality altogether. Did anyone stop to think that even in the midst of so much evidence and credible witnesses, the "proof" seems to always elude us? When you ask the wrong questions, no matter how sound those questions may be, the answers never come.

In the very early years of my research, I started out just like all the other researchers of that time, nuts and bolts research. Just the facts, please, kind of inquiry and don't bother me with all that strange stuff! But problems begin to arise with this approach. No matter how hard I tried to keep my research "pure", anomalous experiences continued to arise in very high percentages of my cases. These "anomalies" were persistent, were reported across the board, and no matter what kind of scientific inquiry I threw at them, they just wouldn't go away.

A single case changed the way I viewed and then investigated the UFO mystery forever. The case involved a group of people star watching on a beach. There were about 14 individuals within about 25 feet of each other. At some point, about 4 of those people witnessed a pretty amazing UFO sighting. When questioned alone, they all described the same thing. They were articulate and in agreement about the shape, lights, and sound. What was extraordinary was the fact that the other 10 witnesses saw and heard nothing. No matter how hard the small group tried to point out the object, the others just could not see it.

This was a revelation to say the least. It dawned on me that there was "something" very different about this small group of witnesses that enabled them to "see". And, bordering on obsession, I was determined to find out what that "something" was. So, being young and poor and without the resources to purchase fancy equipment for research, I did have a steady stream of witnesses that I could study at close range and for long periods of time.

As the years passed and the number of witnesses that I had interviewed increased, other paranormal and high strange events continued to be reported among my UFO witnesses. And…it just would not go away. This trend has held for 35+ years of research, especially with my witnesses who had reported continuing sightings and/or interactions with the phenomena. Furthermore, I found that many other so-called "serious researchers" would edit out those experiences or just discard the whole case. They ended up categorizing those cases under various mental disorders, hallucinations, hoaxes, and in many instances, knew there was really something going on but did not want to risk ridicule for including events that did not fit neatly into our current paradigms.

My research just did not lend itself to these explanations, and I refused to leave out pertinent information just because it might ruffle some people's feathers. I found high percentages of my cases were told by sane, stable, grounded, productive, articulate, and integrated individuals. They were hesitant to come forward or even give their names, wanted no notoriety or attention, were afraid of being labeled "crazy", and just wanted someone to hear their stories and actually believe them. And the cases just kept piling up… So, 35 years later, my research has brought me to these conclusions:

1) UFO Sightings, Entity Interactions, Cryptid Sightings, and Paranormal Events were happening way too often within the

general research to be ignored or discarded. This includes cross-over events (events that report multiple types of high strange encounters, i.e., UFO sightings that include orbs, cryptids, and other paranormal interactions). My definition of Paranormal: Any event, encounter, or sighting that does not fit the standard, current scientific model.

2) A very high percentage of witnesses had valid credibility and portrayed no characteristics of emotional or mental aberrations.

3) Some very strange things were being reported on a regular basis, with consistent content throughout investigated events, regardless of cultural factors, sex, or upbringing.

4) There were repeated patterns, common denominators, and high strange events arising in a very high percentage of cases being reported.

5) Witnesses reported "sensitivities" from birth or developed said "sensitivities" after experiencing a UFO encounter or other Paranormal Event.

6) Previously, there had been much investigation into the events themselves, but very little research had been done in studying the **witnesses themselves**.

7) And the most important: Individuals encountering this phenomenon were forever changed. Witnesses that did not go into denial and stayed the course experienced incredible transformative events. They displayed increased psychic/intuitive abilities, intellectual gains, cognitive increases, and overall general increased levels of maturity.

In this preliminary work, I hope to be able to impart the basic insights of my journey thus far into the UFO mystery and share some truly transformative stories into this virtually unexplored part of UFO research. This work will not be written as "proof", but it is founded on the experiences of very real, credible people and their journeys through some very exhilarating and terrifying events. When new investigators or just the curious ask me about how to get started in their own inquiries, I ask them this one very important question, and I love the way Thom Powell phrased it: "Well, you should decide what you want out of your 'research'. Do you want understanding or do you want proof? They're sort of mutually exclusive. If you chase after proof, you may or may not get it, but you're also going to find that it distances you from understanding and inspiration". "Understanding and inspiration" has been my fire and driving force all of these years, and as this mystery entered many of my witnesses' lives, they were consumed with it also.

As we go forth, I hope to introduce you not only to some incredible people and their brave journeys but give you a glimpse into the mysterious nature of your own true self. This work cannot be spoon fed and it is not for the faint of heart. It is not another new age metaphysical liturgy and is not based on some unattainable intellectual discourse. In some ways, this work can be used as a map of sorts that I hope will take you on an incredible journey. Preparations must be made, companions called upon, weapons and shields forged, bodies fortified, and souls enticed. For once begun and a true commitment is made to the cause, it's as if this adventure takes on life of its own, not only for my brave witnesses but for all that go along on the ride. Instead of clinging to the shore, we let go bravely. We strike out, with courage and curiosity, and also with shaky legs. We let go of our known, safe home and begin to let the river of this mystery carry us forward. I believe all are called in this quest, but only the few will hear that call and venture forth.

This is a collection of stories of the "Few" that did set forth on that brave journey. Who risked ridicule, rejection, and actual persecution for sharing their encounters with the usually hidden and mysterious side of reality at large.

This reality, this life, is a mixture of beauty and horror, pain and ecstasy. It seems you cannot have one without the other. But, curiously enough, that's the blessing. In so many of our belief structures, we are taught to reject at all costs the darker, mysterious, hidden side of creation. We are urged to hold on to that safe spot and reject any value that this hidden mysterious side has to offer.

I feel this rejection of the not so pleasant hidden parts of each of us and our world continues to hold us bound. It limits our potential and holds us slaves to unknown masters. When the human soul is liberated by whatever method and all influences of our complex nature are given their due, a more mature, creative, and happy individual emerges.

I feel these extraordinary individuals who have entered the UFO/Alien mystery and traveled back to us from distant shores have brought back with them something overlooked and yet priceless. If I succeed, I hope to convey to all that take up the journey with me the secrets that have been brought back from these other worlds, many times at a great price. Secrets that, once revealed, may give us the courage to face our own unique journeys, to endure our own transformations, and finally be welcomed into a larger, more wondrous world that has been waiting for us, I feel, for a very long time.

Angelia Sheer, UFOgirl

1. MILITARY SIGHTINGS

"Those with less curiosity or ambition just mumble that God works in mysterious ways. I intend to catch him in the act."
                                        - Damien Echols, Life After Death

Writing "UFO Encounters" has been an adventure and an education in itself. I'm one of those structured, analytical kind of people that likes to place one foot in front of the other but the book just refused to be written in that kind of way. Just when I thought I had everything in the right order, another interesting case would appear that I thought my readers would enjoy and I had to then go about reordering the whole book. At the last minute, a very credible witness appeared that I could just not leave out. How he appeared is just another example of the mysterious nature of this whole subject and what better way to kick off the start of your reading adventure than to share his incredible story.

Being the State Director and Chief Investigator of MUFON of Tennessee, it is part of my job and that of my Assistant State Director's, Josh Cross, to schedule quarterly meetings for our area. Our groups usually range from fifteen to fifty curious individuals and I always look forward to sharing new cases and meeting new people that are interested in the UFO Phenomena. Our final meeting of 2019 was near the holidays so we had a comparatively small turnout due to all of the Christmas bustle, so it was easy for everyone to talk with each other before the presentation. We always wish for large turnouts but I have learned that a small turnout at times can lead to a more trusting, intimate setting. In these settings, individuals are more prone to come forward with long hidden encounters and that's just what happened this meeting.

We were very fortunate that day to have a great lecture from a very credible witness. So, as the meeting came to a close many attendees stayed to mingle, share stories, ask questions and just enjoy spending time with other like-minded people. As we had to vacate the meeting room soon, I suggested to the small group that lingered, that we all meet at a local restaurant and continue our conversation. My first witness and his fiancé were part of this group and this is where our adventure began.

Since our original meeting, this witness, as well as so many others, have worked closely with me on documenting his sighting via this book, searching for other living witnesses, and creating a lasting video reference of his incredible experience. Check out my YouTube channel at "Angelia Sheer" for Ron's testimony and other UFO Encounters told by honest, real people. Like so many of my other witnesses, due to long hours of joint research, he and his significant other have become good friends and I look forward to discussing the mysteries surrounding high strange events such as these long into the future. As I have found with so many of my cases, our first witness is of outstanding character and I want to stress the bravery and determination of his decision to openly share his sighting. At every turn, he has agreed to share his identity, to stand by his testimony and stress the need for open disclosure. This book is a compilation of that kind of tenacity, bravery and determination to go where many will not tread and share with you, my audience, a peek into the mysterious UFO world. Now get comfortable, ramp up your curiosity and dig into some of my best cases over the last 35+ years of research… I think you're in for a wild ride!

Case: No One Saw A Thing: UFO at Sea
Parasheer Research Case Files
Investigator: Angelia Sheer
Location: Atlantic Ocean
Date of Event: June 1970
Witness: Ronald Greene

As Ron and his fiancé entered the meeting room and got seated for our final MUFON gathering of 2019, immediately, my UFO spider sense went off. Let me explain…years ago when I was just beginning my UFO research, I was poor and really had no way to advertise myself. Face book didn't exist and we didn't even have cell phones back then so I would attend UFO/paranormal meetings and then just watch the crowd that attended. Many times, individuals that sat in the back, asked no questions, and attempted to be as invisible as possible, would turn out later to have experienced a sighting. My strategy would be to wait around till the end of the meeting, approach these individuals, introduce myself and my work, and hope for the best.

Many times I was rewarded with a new sighting, a very relieved witness and possibly a new friend! This system worked so well throughout the years, I still watch the crowds and I've gotten pretty good at recognizing those that have been holding on to something for a long time. As I watched Ron throughout the meeting, I was pretty sure he might have a story to tell, so I made a special effort to talk with him before he left. He was polite and his fiancé engaging, so I invited them to join the gang for an early dinner after the meeting.

So off we all go, our speaker for the day, a few other attendees, my team, and Ron and his fiancé to a local restaurant to talk about the lecture and just enjoy each other's company and common interest. I've found credible people like Ron; check you out for a bit (as they should) before deciding to share their stories and as luck would have it, that's just what happened. We were all happily engaged with dinner and conversation, when Ron all of a sudden shared the following incredible account.

Ron's Account:

"I was in the U.S. Navy in June 1970 in the north Atlantic on resupply trip for nuclear subs. My assignment was the weapons department and we steered the ship (The USS Orion, AS.18). I was on my watch in the wheel house from midnight till 2:00 a.m. At approximately 12:45 a.m. one of our spotters radioed in that a spot light from a plane was getting low to our group. The Officer of the Watch told him to keep an eye on it and asked the exact location of the incoming object. Minutes later all called in the location behind our ship at about a distance of two miles back.

It was seen by everyone on watch and the officer on duty had our escort ships alerted about what was thought to be an incoming airplane. The object was not going fast as is it came past our ship and no sound was heard somewhat like a glider of some kind. With binoculars we observed at approximately 200-300 yards off our starboard side, something with dim red lights inside and a spotlight on the front bottom area of the craft. With the binoculars we determined that it was no known plane and we could actually see two figures in what appeared to be a cockpit illuminated in a dull red light.

All details were written down in the log book as is standard procedure. The ocean was dead calm and we could see the thing by the reflection of the water. This thing was now about a mile ahead of us when the red

4

lights went out and green lights came on. It then went from approximately fifty feet off the water to diving down into the water. Our escort ships had spotted the object also and were all talking about the thing submerging under the water in plain view of all of our ships. The submarine that was part of our fleet also radioed that they had watched the object on sonar until it moved out of range at a great speed.

I changed to the next watch at 2:00 a.m. and at that time half the ship was awake due to the strange sighting. Shortly after 3:00 a.m., we heard that "it" was back. I was told by the man on watch, that it came out of the water about 300 yards behind our escort, the interior light went from green to red and it then ascended to about 100 feet over the water. We were headed north moving about 6 knots per hour and the object stayed with us for approximately 30 minutes. Eventually the object rose to a few hundred feet, paused a few minutes, then it picked up speed, ascended rapidly and was then gone from sight.

Later that day our captain told us that all log book notes were gone and we were not to talk about the incident among ourselves at all. We all did as we were told and no one talked about the event openly. It was hinted that it might be one of ours. Later, we found out that one of the men on our escort ship literally freaked out, had to be restrained and was removed from the ship two days later by helicopter. We all witnessed the helicopter land on the ship near Norfolk, Virginia and then leave about fifteen minutes later. The next day we were ordered to disconnect all monitoring equipment on the nuclear warheads and no explanation was given for this command. I estimate that over 100 men on board our ship, the two destroyers and the submarine witnessed this strange event.

Fast forward… in 1983 I was working at (omitted for privacy), during break one night a new employee with a nick name of "spaceman" was talking about a sighting he had back in the Navy. As I listened it

5

sounded like the same event that I had been present for. I spoke up and filled in some of the blanks in the story and he immediately wanted to know how I knew that. I told him I was on the sub-tender. He told me he was the man that was removed from the ship by helicopter as we came into port and was given a section eight discharge. He constantly talked about aliens and was let go from the job two weeks later." -- Ron Greene

You really can't make this stuff up... one of my now favorite sayings! I was so taken with Ron's testimony, I invited him to the house so I could video his story personally. I asked him if I could share his story and he said, "I'm in my 70's now and people need to know what's going on and what a lot of us have seen... I don't care what anyone says!" As we talked about his case, I asked him in his estimate how many other servicemen might have seen the object and he estimated at least a hundred. That's a lot of witnesses out there that we may be able to find. As we speak, Ron has taken action to see if he can locate some of his fellow servicemen and if they might be willing to come forward to tell their stories.

Following are the original sketches that Ron sent to me as well as a picture of the sub-tender ship he was assigned to.

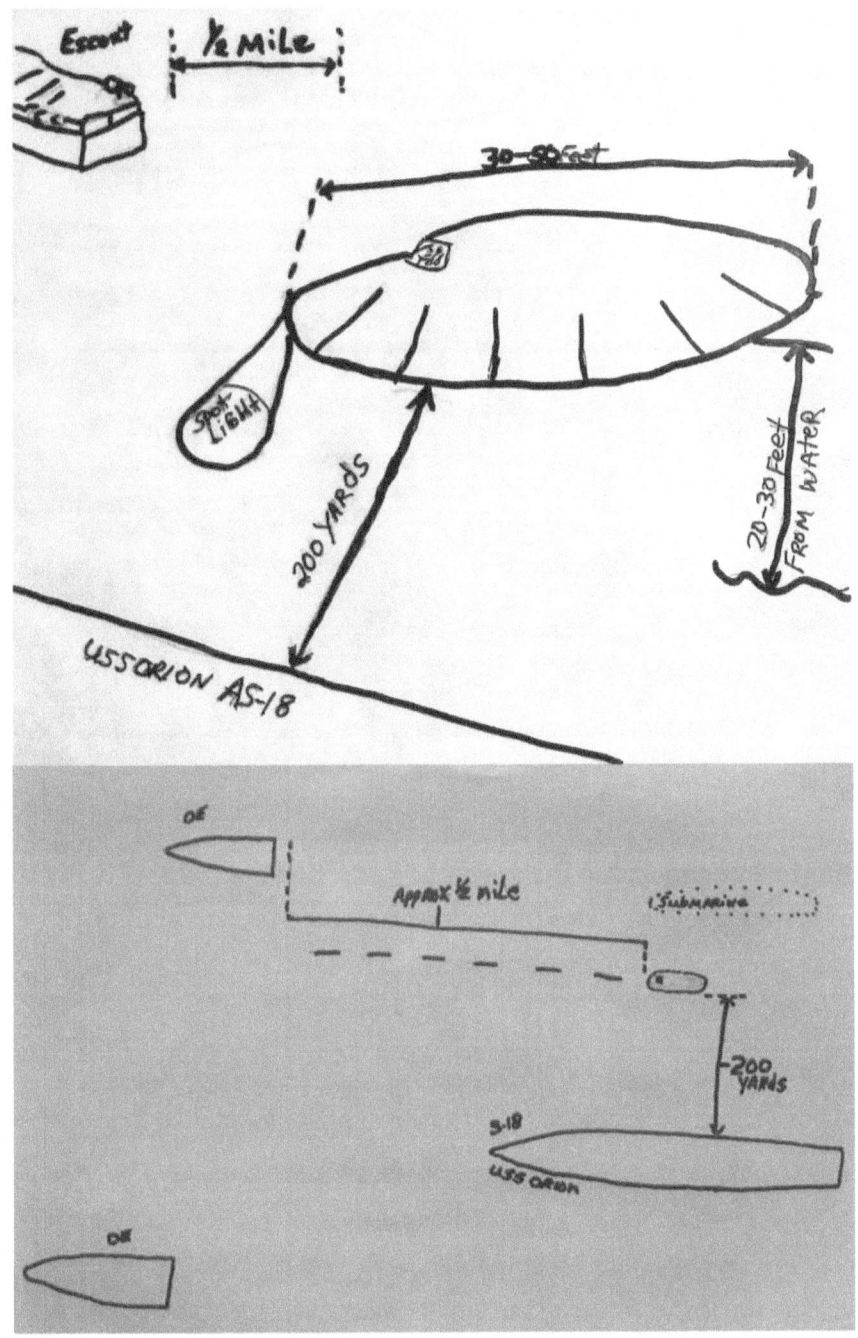

Drawings Courtesy of Ron Greene

USS Orion. Photo Courtesy of Ron Greene.

As we were finalizing Ron's report, I asked him as I do with all of my witnesses how this sighting and any other earlier sightings, (pay attention to the fact that there were earlier sightings) had affected him. Here's his response:

"When I was 11 years old the Russian Sputnik was flying over and we could actually see it in our neighborhood. Other kids were playing outside and we all saw it flying over. As we watched a ball of light flew up beside the satellite and tracked along with it. We all watched for a bit and then went in and told our parents about it which they dismissed as another satellite. I really never bought that!

When I was 13 years old I had the opportunity to visit Oak Ridge with the Nashville Boys Club. While we talked with one of the engineers on staff, he told us that they had a UFO in the "Big Building". We all laughed about that comment but we really didn't know if he was kidding or not and I have now often wondered about his comment.

In 1996 my best friend, his daughter and I were watching for the space shuttle to pass overhead. My friend's daughter had been given a school assignment to watch for the satellite and make a report. I was at my home and we were both talking on the phone as we all watched. As the shuttle came into view a ball of light came into view at a high rate of speed made a 90 degree turn toward the shuttle and again tracked alongside for a period of time. The shuttle has to reach speeds of up to 28,000 kilometers per hour to remain in orbit so do the math on that! We excitedly talked about what happened and what the object could have possibly been. Just this year, 2019, The History Channel showed NASA satellite videos and in one of the captures was the 1996 shuttle with the strange object appearing.

Since childhood I have always been interested in the possibility of UFO's but after my sighting in the military, I guess you can say I became a little obsessed with studying the phenomena. To date I read and watch everything I can find about UFO's and other related subjects. I always watch the skies these days and often wonder what other worlds and beings may be living out there and if what I saw may be a visit from them" -Ron Greene

"A little obsessed"…that's an understatement. As you read along through the cases in this book really pay attention to the sentiments shared by each witness. As we go forward and examine each case in this book, a pattern arises of self-introspection, a questioning of reality, intense study, a struggle with fear, and then finally an overall transformation of perception.

The only cases that this did not happen were witnesses that chose to deny their experiences, avoided any struggle of grappling with the unknown and thus prevented the culmination of a transformative event. This pattern of "transformation of perception" will repeat itself through each case presented (again, except those that chose denial) and can announce itself in a gradual subtle fashion or as most of these cases reveal, carry a person forward on a radical journey they never could have imagined. That journey and its lasting mark upon my witness's lives and perceptions is the truly remarkable part of this mystery and their stories deserve the closest scrutiny.

2. STRANGE ENTITIES

Two possibilities exist: Either we are alone in the Universe or we are not. Both are equally terrifying." — *Arthur C. Clarke*

I love cases that are up close and personal. We can speculate all night about little lights in the sky, but when you come face to face with something strange, it's very hard to dismiss the event. These kinds of sightings change people forever. I have so many that I speak with that say over and over that their lives will never be the same again. Their world views have been completely destroyed, and that kind of change can severely disrupt people's lives.

Many witnesses, when telling their stories, actually break down and cry, men and women alike. They apologize over and over again for their behavior and ask me if I think they are crazy. Each time, I assure them that after 35 years of research, I know "crazy", and that they are not that. It's a testament to the rigidity of our cultural views that these witnesses feel that they cannot share such a life-changing event with the ones closest to them. Many have said to me that it's okay if they are crazy, "because they have a pill for that". This sentiment has been shared with me so many times that it's not unusual to hear anymore. My dear friend "Kathie Davis" from Budd Hopkins' work, "Intruders", is quoted as saying this many times over.

What an awful predicament to be placed in. These every day, loving, honest people have had, in many cases, experienced a life-changing, traumatizing event, and they have no one to tell. They are left alone to deal with the fear and anxiety, and their incredible stories go unheard and, worst of all, unstudied. When I assure them that they are not alone, that there are many others who have endured similar events, they again cry in relief. I have been honored to not only document the stories of these individuals but also share in their struggles as they try and integrate their experiences. The incredible transformations that I have witnessed in these people are both humbling and exhilarating. If they stay the course and work through the fear and uncertainty, I'm amazed over and over at the "new" individual that emerges.

Each of the following sightings had a profound impact on the witness. Some actually "disappeared", meaning they changed their contact information and refused to speak about the event again, at least to me. Others stayed the course, dealt with their fears, studied, refused to ignore the situation, and were rewarded with, in some sense, a "brand new life". In other words, they experienced a radical transformation in their lives. This is important as it is a recurring theme throughout all the following cases presented in this book. Pay close attention as we share each witness' story. As we go, I will be weaving together a picture, a pattern of events, and a series of common denominators that lead me to some profound conclusions about the incredible abilities we all share to some degree. At each point, I will leave clues as to those markers in the author's note sections. In the "Tying It All Together" chapter, I will sum up all of these points and present an integrated view of my findings.

For reader clarification, each of the following cases will be presented from my actual investigation logs. These logs present the case name, origination of case, location, and date of event. All

MUFON cases that are referenced are cases that I investigated personally in my capacity as Chief Investigator for the State of Tennessee. All Parasheer Research cases are events that I investigated through my private research group. Names and other identifying information have been changed to protect my witnesses; all other information is presented as the witness presented the case to me. The "events" that I have chosen to share are from everyday people in all walks of life, who were sound of character, credible, and just wanted someone to listen to their stories and actually be believed. The rejection, ridicule, and dismissive attitudes my witnesses are subject to at times is isolating and emotionally destructive. With this work and work from other "boots on the ground investigators", we hope to reach as many other witnesses as we can find and provide a safe place for their stories to be told.

So, without further ado, let's jump into the good stuff!

Case: The Captain and the Protector
Parasheer Research Case Files
Investigator: Angelia Sheer
Location: Dyersburg, Tennessee
Date of Event: Summer of 1973
Witness: Jack Parker

At the time of the sighting, Jack was attending college at The University of Tennessee at Martin during the day, and in the evening, he worked the night shift as a police officer. Sometime around 2:00 a.m. on a day in the summer of 1973, the witness was sitting outside the police station in his patrol car in Obion, Tennessee, when he spotted a V-shaped craft. He turned up his spotlight in the squad car, then proceeded to get into the car and drive after the object. At this point, the witness estimates the craft to be about 1000 feet in altitude. As he proceeded to watch for the craft, he noticed a light that appeared similar to a search light back toward town. At that point, he called Union City to report a low flying aircraft, as he was concerned that a crash may be imminent since the object did not make a sound on its current course. Jack then proceeded straight east away from the police department as the object had never made it to town. The witness continued in pursuit of the object, noticing it was scanning the pastures and road with a high beam light as it drifted along. At that point, the object was about 100 feet in altitude and seemed to be descending as it went along. As Jack approached the object, he was able to again turn his spotlight upward and could clearly see the object in question. It was definitely V-shaped, had slowed considerably, and, in a surprising maneuver, had turned and passed over his patrol car to land in a nearby field.

Jack was sure this was no normal aircraft, and as he brought his patrol car to a stop, he reported being afraid to step out of the car as he was sure he might die. He turned the patrol car spotlight toward the landed craft once again and estimated that it was about 100 feet away from the patrol car. The object was emitting a very bright light and a slight humming sound was clearly audible. Not knowing what else to do, he then called over to Union City to report what was happening. The only memory he has of that call is the dispatcher saying they would send out a rescue squad.

The next memory Jack was aware of was of a purple light that began to engulf the patrol car. The other officer in the passenger side of the car was hit completely with the light, and he seemed to go into some kind of trance-like state. The purple light then hit Jack's arm, as he had his elbow sitting on the car door out the window, and it immediately became numb. There is an extended break in the witness' memory at this point. When he came to, there were no lights on in the patrol car, and he had no idea where he was. The next thing he remembered is his patrol car being on a railroad track, and he was almost hit by a train. (Extraordinarily, this happens to another family later on.) When they did find their way back to the patrol station, it was about 6:00 a.m. His last memory was around 2:00 a.m. The witness feels that there must be a recorded conversation with Union City as he really thought at first an aircraft was going to crash.

As life will have it, time passed for Jack and his family, and the troubling experience that he had endured was buried for a time. He had a family to care for and his career to consider, so the missing memories faded under the normal pressures of life, and Jack just pushed on. It was at this time that he took a position with the Dyer County Sheriff's Department and came across an advertisement for a smoking cessation class. He had tried to quit smoking numerous times before without luck, so he decided to attend the meeting which was being held at a local hotel conference room. The "class" turned out to be a group hypnosis session to not only help people stop smoking but to also lose weight and/or cease other unwanted habits. Jack reported that he was excited about the prospect of quitting smoking, so he settled in to begin the session with the other attendees.

As soon as the session began, Jack reported feeling "strange" immediately. He began to feel disoriented, and then all of a sudden, he experienced being back in his patrol car so long ago on that fateful night. It was then that memories of what fully happened that night began to pour back into his mind. He was once again sitting in his patrol car seeing that strange craft fly over, land, and then witness that purple light engulf the car and his partner. He relives the fear completely as the memories continue to pour back into his mind's eye. He remembered looking in the mirror and seeing 3 small beings approach the driver's side of the car. As he is reliving the incident, he says that he really wasn't that scared since he had his .45 caliber with him. The beings continued to approach the car, they opened the door, and they took him by the arms. The young officer that was in the passenger side of the car was also taken, and he seemed to be really out of it. For some reason, Jack still felt very clear about what was happening to him (unlike his partner), so he could observe his surroundings and the beings taking them while he was being led over to the craft.

He remembers that the entities who took him were small like children, and the witness described them like the "Grays" that you hear about. Jack describes that two of the beings had him by his right arm, and one being was holding his left arm. The other officer was being led in front of him, so the witness could observe what was happening to him also. They all proceeded along in a deliberate manner until they were all inside the craft, at which point he remembers asking, "What's going on?" The beings who were escorting him answered inside his head, "We just want some information and you will not be harmed." Jack then immediately asked, "Why can't we just talk?" It was at this time another being that was standing up on a deck near what appeared to be a control panel instructed the others to release him. He motioned for Jack to come up the ramp and over to where he was standing. As Jack approached this other being, he felt that this being was somehow in charge, and Jack innocently asked him, "Am I dreaming?" The witness eventually named this being that was in charge the "Captain" and interestingly, the "Captain" called Jack "Protector".

The Captain told Jack that he was not dreaming and then kindly started showing him the console and began to share information about his people, the earth, humans, other aliens, and other pertinent information, seemingly via telepathy.

The witness reports that the Captain spoke to him in his head and told him the following. He said his race came to Earth thousands of years ago, but he and his people felt like this was now their home also. Jack was told that the Captain's people like the water and live in underground bases in the middle of the Pacific Ocean. The Captain showed Jack pictures on the console of Washington State, and their base was approximately midway out in the Pacific Ocean from other land masses. Jack was told this was a safe haven for the Captain's people and that the Captain was born there. The Captain also communicated that he did not know his parents like humans do.

The Captain also told Jack that many other races visit Earth. Some just explore, others are not so kind, and his people had been protecting Earth from some of these other marauder races. He told Jack that they also studied humans along with other peaceful races and did not understand why we were so violent. The Captain said at this time, humans were not allowed to go into space because many other races fear us. In the past, there had been terrible wars on earth over our natural resources and other issues, and in general, the greater civilizations at large did not want that violence spreading out into their realms. The Captain also shared that, in the past, certain ancient aliens modified early humans with their own DNA and that humans actually took on some of their violent and aggressive tendencies. He expressed that they are still trying to help with that because they like humanity and want us to take our rightful place in the universe and join all of the other races that exist out there. Jack then asked about the engines on such a ship as he was not aware of any sound, and he was told that it was some kind of antimagnetic drive. The Captain was very pleasant to him, and he never felt threatened by this being. He described the Captain as wearing a tight-fitting uniform that was slightly metallic looking in nature and was different than the others in that it had a stripe running diagonally down from the upper left to the lower right side. This stripe was golden in color.

It was at this point in his memory that he happened to turn around and notice a very tall insect-like being examining his fellow officer on some kind of table. Jack was startled by this being and his appearance. After they were finished with the other officer, he was told that the examinations were completed, and they then did something to Jack that knocked him out. It was at this time that Jack was sure they tagged him with a tracking device. Jack reports ongoing visitations throughout the years after his initial encounter.

In one of his ongoing sightings, a huge ship appeared over Jack's house. The witness reports that he was not sure how big it was but estimated that it could have been up to a half a mile in length. When he was brought aboard the ship, that same, kindly being greeted him. Jack commented that "the Captain had certainly moved up in the world with this spectacular ship". The Captain then made a comment about the witness' two sons, so Jack felt confident about being tracked and also felt certain that his children were also. After this encounter, there was again some degree of missing memory and/or time. When the witness again regained full consciousness from this visit, he went back into the house and there was "a huge commotion going on!" NASA reported that something had crash landed in their area, and they had sent out a retrieval unit to recover the object. My witness felt that this was not true and just a cover story for all of the people who had seen the large craft that day. This would have been around 1978 or so, when his children were babies. I did a search for this time period, but information was scarce, and I could find no other sightings, reports, or any NASA press releases for these dates or locations. Interestingly, there was a Russian satellite crash in Canada in January of 1978 that scattered radiation over a wide area.

Some years later, the witness had a severe pain in his nose. He started pushing on the painful area and then blew his nose strongly, and an object like a piece of hard crystal came out. The object fell into the sink accompanied by some blood and mucous. Jack could see it sparkle like glass, and it was shaped like an octagon in the middle and tapered on each end down to a point. The object was not collected and actually went down the drain. After that, the visitations ceased and the witness was sure that the object in question had been some kind of tracking device. That would have been in the late 90s. Jack reports that until now he has never shared his story and would like to have hypnosis again to see if he can retrieve more memories of his experiences with these beings and the Captain who he considered to be his friend. He said, "They seemed like a kind people and that they liked humans." It was at this point in our conversation that Jack gave me verbal permission to print his story and vowed to the truth on his father's grave. Jack shared that he went through a divorce at that time, and due to his character and notability in the area, he was given custody of his children. He also ran for public office for constable twice and was elected both times.

I was immediately struck by this witness' sincerity and his need to share his story as soon as possible. He confided in me that he was in poor health and felt that he might not be here for long. He wanted to get the word out that UFOs are real as well as the beings that pilot them. This is just one of many cases where witnesses come forward when they feel they might not live much longer. At this point in their lives, they feel no one can really hurt them, take their pensions, or hurt their families. They just want to relieve themselves of a heavy burden that many have been carrying for a very long time and just have someone believe them. Many have never shared their stories with spouses, family, or friends. Once you have been inducted into the UFO world, it can be a very lonely place.

The witness stressed how these encounters with the "Star People" had changed his life. He wanted to share the following with my readers:

"I no longer think about the world as I did before, and I now know there is so much more to our lives than I have ever imagined or been taught. I am eternally grateful for my relationship with the 'Captain', and even though I know things possibly happened to me that I cannot remember now (physical exams, etc.), I know it was just the Captain's job, and I still consider him to be my friend. I was sad when my interactions with the Star People ended, which coincided with my nose bleed and the possible loss of a tracking implant. I will forever be in wonder over our relationship and am thankful for the other worlds the 'Captain' so graciously showed me."

- Jack Parker

I speak with Jack to this day and worry a little over his health. He has had several close calls but seems to always rally. Every time I speak with him, I'm immediately reminded of his kind nature and honest spirit. Thank you, Jack Parker, for the courage to share your story with the world!

Author's Notes: Things to Remember

1) In numerous encounters, the witness lost consciousness and experienced memory loss.
2) There seemed to be some kind of ongoing tracking of his family as indicated by a conversation between him and the "Captain" who commented on his sons. Witness later describes a nosebleed and discovering what may have been some kind of implanted object that was dislodged from his nose and lost down a drain.
3) A military investigation seemed to transpire around one of his major sightings, and this was publicized as the retrieval of a downed satellite or other terrestrial object.

Case: Late Night Car Encounter
Parasheer Research Case Files
Investigator: Angelia Sheer
Location: Melita, Michigan
Date of Event: Summer of 1981
Witness: "DN"

I can't imagine the fear some of my witnesses must endure! Can you imagine riding along in your car on a pitch-black night on a lone road and have the following encounter happen to you? As you read through the following report, pay attention to the similarities to other cases documented here. Many professionals want to label these accounts as partial dreams or even delusional events. I'm not buying it! Humans do not have completely similar delusions or hallucinations. And, in this case, there were multiple passengers in the car who reported the very same encounter. I spoke with each witness on separate occasions, and each described a similar sequence of events. So, without further ado, let's go deeper into the rabbit hole.

At the time of the sighting, the witness was 24 years old and had been visiting with her parents for the afternoon into the early evening. She remembers the day well as it was a warm summer evening, and it had been fun visiting with her family and nephew. As the festivities drew to a close, the witness reports she left her parents' home around 10:00 p.m. She and her nephew were very close, and he asked if he could accompany her home that evening and stay with her since he was on summer vacation. The witness lived nearby, and the drive home should have taken only about 15 minutes.

"DN" left her parents' house traveling north toward Skidway Lake and traveled about 4 miles up the road, stopping at a stop sign. Her nephew was riding in the back seat of the car. He was touching her hair and remarked to her about how dark it was outside. As the car came to a stop, it all at once felt like some kind of force took hold of the car and pushed it to the side of the road. The witness was at once terrified as she had no control over the car. She immediately realized something very strange was taking place. As the car was forced to the side of the road, it immediately got really bright outside with light flooding into the car and almost blinding them both. Both were immediately frozen in fear and were desperately trying to figure out where the blinding light was coming from. As they were straining their eyes to try and see what was causing the intense light, an entity appeared out of the glare and approached the car. The witnesses were now in a panic and immediately went about making sure the car doors were locked. The entity seemed not in a hurry and took its time while walking around, examining the car and peering in at DN and her nephew.

In comparison to the car, the being could look right in the car window directly at the witnesses without having to bend down to see in. DN described the entity as having no hair, a roundish face, and very dark eyes. Its skin was grayish in color and appeared somewhat translucent, like you could see under the skin. Needless to say at this point, the witnesses' fear was off the charts as the thing did not appear to be human at all. When it walked away, DN reports that both she and her nephew were still frozen in place as they watched the ship depart. They were so traumatized, they both just sat there in the car for a while. When questioned further about how long the two of them sat there, the witness is unclear about how much time elapsed and has no memory of how they got back home or how long it took. For many years, much of the encounter was "blanked out" of the witness' mind, and some of the more disturbing details returned over a long period of time. DN feels that this was some kind of protective mechanism because the whole experience was so terrifying and surreal. As in so many of my cases, the witness has now come forward, years later, because of the emergence of new memories surrounding this event and because she wonders if something else happened.

Here is the witness' sketch of the entity she saw approach the car!

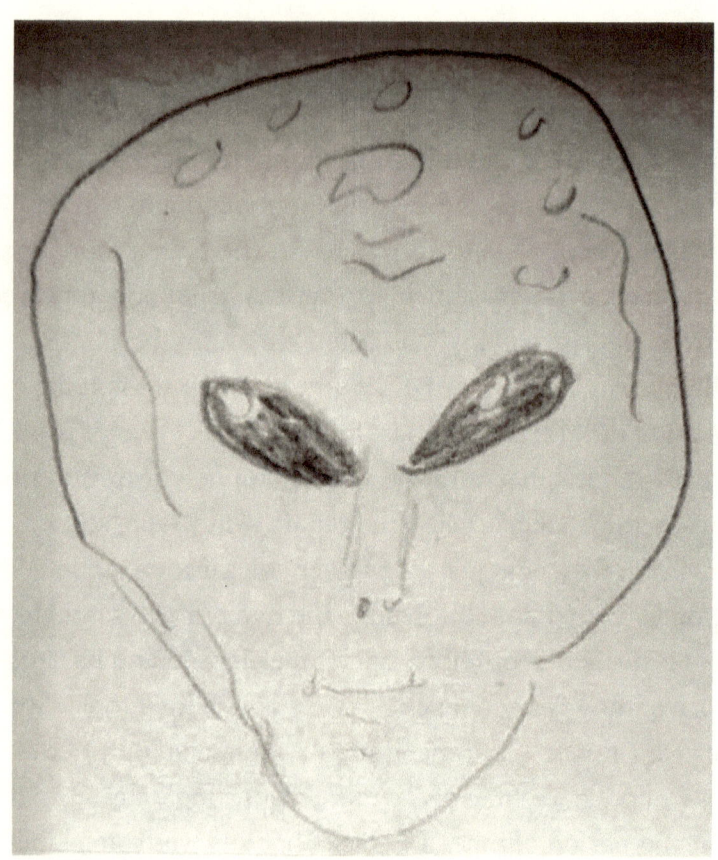

Original Sketch by Witness "DN"

Message from "DN":

"Since that fateful night years ago, when my nephew and I encountered that craft and its occupants on a lonely road in Michigan, my life has never been the same. I often look out at the stars and wonder about all the worlds that are scattered among the stars and the beings that may live there. I have struggled with the memories surrounding this event, and over time, more pieces to the puzzle have returned to me. Just recently in church, a member asked me about my encounter and confided to me that she and her family had witnessed what they thought to be the same object. I was amazed and realized there must be thousands of

individuals out there with similar stories. I told Angelia about this meeting, and I hope to get her together with these folks to compare notes. It is my wish that more people will come forward if they have experienced a sighting, as every voice, every story, builds a foundation for others to feel safe enough to share their own encounters. It is my hope that my story here will be part of a greater picture and will help usher in a greater acceptance for other life and other worlds." - "DN"

I have also had the opportunity to speak with her nephew on a separate occasion, and he described the sequence of events of that night from his perspective. Both describe a similar and terrifying event, and, interestingly, neither can remember how much time elapsed from the time the car was pushed to the side of the road to the time they arrived home later that evening. Again, this is a common theme among witnesses. Many witnesses report that time seems to stand still and that there was this otherworldly feeling that permeated the atmosphere. Witness DN said many times that some memories came back much later in life, and she sometimes wondered if hypnosis may help in recovering more of what happened. To this day, she still ponders on the events of that evening and wonders if she will ever really know for sure.

Author's Notes: Things to Remember

1) The witnesses were driving down a dark road late at night. A common occurrence preceding state changes.
2) Both witnesses reported being frozen in place and that it took some time to be able to move and function normally again. (Symptoms of an actual state change.)
3) The witness reports moments of partial memory of that night and wonders if hypnosis may help her solve that mystery. Memory is "state dependent" and can only be retrieved in many instances if that "particular" state can be recreated.

Case: There's Someone in the Bathroom
MUFON Case Files
Investigator: Angelia Sheer
Location: South Tennessee
Date of Event: January 2016
Witness: "Michael" (Names, locations, and other identifiers were changed to protect the witness.)

Witness Report Paraphrased from Original Report

"I was working the graveyard shift at the time on an important project when the incident occurred. At that time of night, there were very few employees present, and just a few of us had stayed late. Nature called, so I left my workstation to go to the restroom. It's quite a trek through the facility to get to the bathroom, so I headed out down the hall. I had to go down quite a ways, pass through some double doors, and then proceed through a long hallway into the men's room. After using the restroom, I proceeded to wash my hands and then head back to my office. It was at this time that I felt as if I was not alone and that someone was watching me. Startled, something caught my eye, and I turned around to notice someone wearing a strange cap that was somewhat flat on the top. At first, I thought I was seeing things. The "man" seemed to tip his head to view me from under the brim of his cap, trying to keep an eye on me, and I immediately moved over to the wall so I could see the "man" better and put some distance between us.

The "man" had bright, pale skin with gray tones and was wearing pants and a jacket that seemed to be made out of some kind of shiny silver material. His shoes were somewhat normal-looking, and he was wearing some kind of shades over his eyes. I could make out his eyes under the shades and as I observed him, his

26

eyes did not appear to blink. The individual quickly turned around to face the wall when he realized that I had noticed him, and he appeared to be quite nervous and shaking. At this point, I was only a few feet away from him and began to realize something was just not right about this whole situation.

I spoke up and said, "I didn't mean to startle you, and I'm about to leave so you can have it to yourself." I turned around to finish washing my hands, and even though this was a bit strange, the overall impact of what was happening had not really sunk in. I was aware the "man" kept turning around to look at me, and I actually felt sorry for him. I wondered why I had never seen him before and just mentally thought that maybe he was a new employee and was just nervous about his first night on the job.

It also dawned on me that I never heard the doors open to the bathroom, and I was sure that there was no one in the room when I entered. At this point, another employee starts to open the bathroom door and the "man" looks more nervous than ever. He quickly swings around and starts to walk toward the back wall at the end of the stalls. Shockingly, a strange light had begun to appear brighter and brighter at the wall, and the "man" ducks his head and literally leaps into the light through the wall.

I was shocked to say the least and frozen in place. The other employee was fully in the room at this point and proceeded to do his business. The presence of the other employee helped break me out of my shock and I turned, shaking, and walked out, not knowing what to think or do.

In retrospect, I'm sure the "man" was a real solid being. I could hear him brush against the wall, and his shoes made scuffing noises as he moved. He was extremely nervous, and I realized later that he did not become so nervous until he realized that I could actually see him. I have thought hard and can remember no discerning patches or other identifying badges on his clothing or hat. His face was humanlike but also different. I could see no hair from around his cap, he had a long, thin nose, and I got the impression that his ears were larger and possibly pointed under that cap. It also dawned on me that no new employee would be working that late.

I was hesitant to report this to anyone and have told no one around me, especially anyone at work. I fear for my job and my sanity. I have had no other strange experiences before this and hope that I never do again. I cannot relate how this has affected my life. I cannot sleep, have disturbing dreams, and I think about this event over and over. I really was not afraid of the "man", just the strange, surreal details of the event. I try to look at the event from other perspectives, and I know for sure the being was really afraid that I could see him, and this seemed to be out of the ordinary for him. I wanted to make this report so someone else could know what happened, but I in no way wish to discuss this event as it's just too disturbing."

I did manage to speak with Michael one time as I was fascinated with this case and wanted to try and do an onsite investigation. I have found in the past that most employers will not actually allow said investigations on business property due to legal reasons and the overall stigma that such encounters engender.

The witness said he would think about talking with me again, but he changed all his contact information, and I never was able to reach him again. I was sorry for that as I could hear his real fear and the burden of loneliness that these encounters cause for my

witnesses. I chose to share this case because it embodies the
strangeness of many other entity cases, and it also is a testament
to the effect these encounters have on people and their lives. I
wonder about this witness and hope that someday I may hear
from him again and that at least he has been able to share his
incredible meeting with someone else.

Author's Note: Things to Remember

1) The witness was working the graveyard (overnight) shift.
   Only a few other employees were present at the time, so
   the building was quiet except for the sounds of the
   machines running. This is a perfect time and location for
   individuals to experience state changes.
2) The entity was really surprised that the witness could see
   him! This indicates that possibly the entity had been in
   similar situations with humans before but went
   undetected. What was different about this time and place
   that enabled the witness to "see"?
3) The witness experienced fear and lasting anxiety over his
   encounter. He stated often during our interview that he
   constantly worried they would return for him!

Case: Peeking Aliens, Part 1
MUFON Case Files
Primary Investigator: Angelia Sheer
Field Investigation: MUFON Star Team
Location: Clinton, Tennessee
Date of Event: May-June 1994
Witness: Aaron Hill

At the time of the event, Aaron was approximately 9 years old
and lived with his mom and extended relatives on a family farm.
He and his mom lived in a separate house from the other family

members, but they all shared in the work and convenience of the family property, a beautiful rural area nestled in green rolling hills in east Tennessee. Aaron's mom worked quite a bit, so he was left in the care of his guardians at times, and he enjoyed a happy childhood playing outdoors in the fields and pastures of the family farm. The witness recalls that it was late spring/early summer when he had his experience. The witness remembers it was a warm Saturday evening, and he had been playing outside all day. As was his routine, he showered then went back to a back bedroom to choose some toys that were stored in old milk crates. He remembers choosing the tub of toys that he wanted to play with that evening, grabbing them up, and starting out of his bedroom. As he came out into the hall, he was overcome with a paralyzing fear that something was watching him. The house had three bedrooms with a hall down the middle that led out to the den where his family members were watching TV. As he looked across the hall into the bedroom that was used as a Bible study, he noticed that the blinds were raised up. Although the hall was dark, a streetlight outside the window illuminated the area somewhat, and he could clearly make out shapes in the window.

When he first glanced up, initially, he saw two beings staring into the window. One of these beings looked like a typical Gray, and the witness estimated it to be about 4 feet tall. The next entity was taller, about 5-6 feet tall, and standing to the right of the smaller entity. This taller being had a smaller head and a long skinny neck that was at least 8 inches long. This being would turn its head in a very inhuman manner which only added to the young man's terror. The witness remembers that the being would look very slowly side to side, then it would look to the left in such a way that he felt that it was looking at something else. At that point, another entity appeared in the window, just left of the short

30

one. The witness described this being as also having an unusually long neck.

The witness was frozen in shock and disbelief. But even this he questioned as he was describing the incident to me. He wasn't sure if it was just his fear that held him in place or if the entities were doing something to him. He describes just standing there holding his tub of toys for what seemed like a very long time. At this point, the entity that was the last one to approach pressed its face against the window and just stared in at the witness. Needless to say, this turn of events so heightened this young boy's terror, he was able to break from his frozen posture and fly down the hall to the safety of his grandparents. When he burst into the room, he yelled at (name omitted to protect identity) that there were things staring in the window. Aaron was really terrified, and the people in the room had to have recognized that this was not a normal behavior for him. Now, here is where the odd behavior from the adults comes into play. The witness then describes his family member as calmly getting up, turning the TV up, shutting the blinds, locking the doors, and then just sitting back down to continue to watch TV. The witness can even remember the program playing on the TV which was an Atlanta Braves baseball game.

Late Spring of 94, On a Saturday night

I Remember this one the best

It was the last to appear in the window & was tilting its head back & forth

Original Sketch By Aaron Hill

It is very important to note how strangely Aaron's family member behaved around the possibility of suspected trespassers near the house. This happened in the Deep South where guns are plentiful and no good southerner would ever let anyone encroach on his property. The witness was surprised in the retelling of these events that his guardian didn't even go to look out the window or, better yet, go outside with his gun to check for prowlers. When the witness was older and began to question his strange encounters and the odd behavior of these adults, he was met with complete resistance. The family members refused to talk about any of these experiences and displayed obvious emotional turmoil and denial concerning anything regarding this topic. It was obvious to my young witness that they knew what he was talking about but staunchly avoided any discussion of the event!

Over the years it has become apparent that many Experiencers have a familial history of UFO sightings, missing time, and high strange events. Some families are able to talk about these experiences together, but unfortunately, they are in the minority. A majority of witnesses find themselves in families that have held a secret vigil with their terrifying encounters. These witnesses feel incredibly alone and find if they pursue family members about their experiences, they are shunned, humiliated, and often times ostracized.

Fortunately for this witness, he was able to discuss these events openly with other family members and my team. From the first moment that I spoke with this witness, I was moved by his honesty and sincerity. He confided in me about how this event had really affected him and how it probably was the seat of a deep-seated anxiety he had carried most of his life. He also shared with me that he had been plagued by nosebleeds when he was young, and the nosebleeds had followed him into adulthood but had lessened as time had passed. In my research, as well as the research of other serious investigators, it has been found that nosebleeds are a common occurrence in alleged abductions, missing time, and entity sighting reports. This report of nosebleeds became very important as the night of the investigation rolled on. So after a few initial conversations with my witness, I was so moved by his testimony, I decided to take a team to the area to do an in-person investigation. What would unfold during that trip is what makes UFO investigations both exhilarating in the hunt and mystifying in actual outcomes.

Following is a painting done by Aaron in an attempt to capture the nature of his encounter. I just love this rendering as it is poignant and startling all at the same time. Notice also the reference to owls in the picture. There are a lot of cases of witnesses who report seeing owls looking into their windows. Many in the field feel these reports of owls, deer, etc. staring into windows may be some type of implanted screen memories to cover for what was actually being seen. For an in-depth look at this phenomena, check out the work of Mike Clelland and his book, "The Messengers: Owls, Synchronicity and the UFO Abductee". Very interesting read!

Original Artwork by Aaron Hill

MUFON Field Investigation

Location: Clinton, Tennessee
Date of Event: July 27, 2018
Team Present: Angelia Sheer, SD, CFI, MUFON Star Team; Josh Cross, FI, Assistant State Director, MUFON Star Team; and Don Williams, FI, MUFON Star Team

Even though this case was considered a historical case at the time, I was so taken by the witness' sincerity and his profound story, I decided to take a team to the area and just see what we might find. When we arrived, the witness introduced us to his mother and other interested family members, and we all just got to know each other for a bit. I can't stress how important this is! Many of my witnesses have not only experienced and seen things that have completely shattered their world views, they have taken on fears and anxieties that were non-existent before their events. Trust plays a major role in helping the witness to feel safe in discussing what has happened to them. Many times, more trauma is accumulated when the witness tries to discuss what happened to them but is met with ridicule and disdain.

After our initial discussions, general walk through of the farm, and completion of necessary paperwork, the team and I decided to explore the general area, and then we would come back to the farm later in the day toward nightfall. We found another report of a young man approximately 16 years of age (remember this was back in 1994 and the laws surrounding fishing close to the steam plant were different. I do not encourage anyone to investigate this area without proper knowledge of the current laws that now protect strategic sites that produce energy, provide hubs for communication or other sensitive areas) who reported seeing similar creatures near The Bull Run Fossil Plant on the Clinch River. Bull Run is a steam-generated power plant managed by the

TVA. At the time of the incident, the witness was fishing in an estuary that the plant uses to dump water used for their steam production. When the water is returned, it is somewhat warm, and the fish seem to congregate in those warmer waters. In construction of the plant, the Army Corps of Engineers built a strip of land that juts off the bank where the water is returned, and it is on this land the young man was fishing. On the evening in question, the young witness was walking out onto the manmade peninsula around 8:00 p.m. to fish when something caught his eye. In front of him were 3 alien-like beings with long necks just standing there up along the path. One of the beings actually looked at him and pointed. To say the least, this was all it took for him to hightail it out of there. The witness immediately called his sister who described him as almost hysterical and crying. We could never get a personal interview with this witness, so I must leave this report as unsubstantiated, but it's beyond chance, at least in my view, for such a similar report to arise around the same time frame. The description of the young man's distress was extreme; I was sorry I could not locate him and just have a conversation about his experience. Many times, sharing your story with someone who understands the trauma these events can cause can actually relieve the individual of some of the anxiety they carry.

We did visit the Clinch River site where we walked along the river footpath and took some photos of the area. We questioned some local folks about the area, but no one had anything to report, and we found nothing unusual that may have aided in the investigation. After finishing our investigation at this location, we headed over for some dinner and then returned to the original location to start our major investigation of the property. Little did we know that the events that would transpire later while we were on the property, would continue to validate the many baffling phenomena that surround the UFO mystery.

Photo by Josh Cross along the Clinch River taken by our drone: From left: MUFON STAR Team Members: Josh Cross, TN Assistant State Director, Don Williams, Angelia Sheer, State Director, Chief Field Investigator

After we finished our investigation at the Clinch River site, we returned to the original location to continue our primary investigation. As per our normal procedures and while we still had plenty of daylight, we started our survey of the property. We had studied maps of the area before arriving, so we had a pretty good idea of how we wanted to proceed with the investigation. I started my general line of questioning with the witness and his family while my co-investigators started their sweeps. Mr. Don Williams is one of my primary team members coming to MUFON with long experience as a pilot with a military background. Don has earned STAR Team membership with many hard hours of work. He has accompanied me on many field investigations and been a major asset in the functioning of the group. He provides transport for heavy equipment such as 4-wheelers and other major equipment we may need for long investigations. Don has also been present on some major sightings we have had as a group, and I was very glad he was able to accompany us on this investigation. As I started my conversation with the primary witness, I noticed Don was off already questioning family members and friends about the local activity.

Also present on this trip was my Assistant State Director, Mr. Josh Cross. Josh is a firefighter/paramedic, and at this point, I have lost count on the lives he has saved. His knowledge, positive attitude, and the ability to think outside the box has been invaluable to our core team and has also earned him a place on our STAR Team. On this day, Josh had gathered up the electromagnetic (EM) field meters as well as our "Bug Detectors" and set out to start his sweeps of the property. The first thing he does is test the property for any high magnetic field readings. High EM fields can affect human beings in unsettling ways. Long exposure can cause headaches, dizziness, and sometimes even hallucinations. We always start out with these readings as a baseline and to rule out human perceptual anomalies. As Josh was making his way around the primary house where the sightings occurred, a strange thing happened. While measuring with the EM meter, the "Bug Detector" in his pocket starting going off around a particular tree. To say the least, that is very strange. The "Bug Detector" is used inside to test for listening devices and is calibrated to the 3 main cellular companies with very specific frequencies. A tree or any living thing should not set it off. Josh, thinking that he had just not adjusted the calibration correctly, reset the detector and went to place it back in his pocket when it went off again. At that point, he called me over to take a look, and no matter how we tried to recalibrate this piece of equipment, it just kept going off like crazy. All in all, we found 3 trees that were in a triangular pattern around the house where the entity sightings were that set off the detector. This constituted one of the first of several truly baffling instrument readings that we would encounter throughout the night. We did record some magnetic field readings that were on the normal side of the spectrum, except they did not fluctuate and held steady during our investigation. Also, the stranger thing that we recorded is the off-the-scale reading of our "Bug Detector" on several trees on the property as mentioned above. This radio frequency detector is used to check for bugging devices and usually only used in a home or other structure. It was by complete chance that we discovered the readings surrounding the trees. In 30 years of research, I had never seen that before. And, as the night rolled on, things just got stranger and stranger.

After making note of the anomalous readings, we proceeded on to complete our sweep of the property and prepared for sunset. At this time, we decided to send up our drone to take some estimates of distance concerning other lights that had been spotted in the hills across the street, so we ended the property investigation and prepared the drone to go up. It was a nice night, and we all had a great time talking with the witness, friends, and family about the case. The drone functioned well, and we were able to make some estimates about some orbs of light that were spotted skimming the treetops across the way along a hilltop. The night flew on by, and as it was getting past 11:00 p.m., we started packing up all of the gear, saying goodbyes, and making our preparations to leave. I was in the house speaking with the witness' mom and thanking her for allowing us access to the property when Josh came rushing in and excitedly asked me to come outside. When I got outside, the witness and Don were standing under a security light next to our vehicles. Josh exclaimed that he had an idea he wanted to try with the "Bug Detector". Remember, we usually only use this device for sweeping homes and buildings, but due to the strange readings on the trees earlier, he had an idea to try it out on the right side of our witness' nose where he had reported so many nosebleeds. I thought this was an interesting idea, so with Aaron's consent, I told Josh to go ahead and give it a try.

On the first pass near Aaron's nose, the thing went off like crazy. We stopped, recalibrated, and tested on other team members as a baseline with no activity reported. Again, as Josh brought the detector close to the witness' nose, it went off to the max. We were all stunned! In 35 years of researching, I had never seen anything like that. Josh was anxious to try again, so he once again recalibrated and placed the instrument up to Aaron's face. Again, the meter went off, but this time, the witness' nose began to bleed profusely. We were filming this segment of the investigation, and the nosebleed was so bad, we were forced to stop filming and get towels to try and stop the bleeding. It was at this moment that the security light above our head went completely out for no apparent reason. I remember my witness standing there with a towel pressed to his nose, exclaiming, "Well...I guess we stirred up the hornets' nest." You really just can't make this stuff up! We were all floored and a little spooked!

We repeated our procedure one more time with Aaron (with the same results), tested each other for a control, and then spent some time calming my witness and his family. When we were finally packed up and ready to go, we had about a 4-hour trip home, and I have to admit we were all a little spooked the entire way back.

Photo by Angelia Sheer with permission by Aaron Hill

More Strangeness

In the days following our investigation, we all started to receive numerous strange phone calls. With one of mine, I was getting into the car when my phone rang. I let it go to voicemail thinking I would return the call hands free when I got settled in the car and was driving. When I went to call the number back, it said it was disconnected and no longer in use. How could the number be disconnected and no longer in use when I had just received a call just minutes before? This happened to me, one of my fellow investigators (who was out of state), and to the witness and one of his family members. We also received weird calls that, when returned, indicated they were from a military base of some kind. Also, my witness called me late one night, very agitated, stating he felt he was being followed by a black SUV. I told him to pull into a police station or other high-traffic area and call me back immediately. I didn't hear from him for a bit, and I have to say, I was very worried for his safety. Luckily, nothing happened, and he got home safe and sound. Unfortunately, these kinds of events surround many of my and other investigators' cases, past and present. "Men in Black", black SUVs, strange phone calls, and helicopter flyovers seem to be part of the ongoing mystery that shadows so many UFO and high strange events. As I was finishing up writing this report for the book, another case came in that was eerily similar in its telling. Now take a look at just how strange this field can be.

Author's Note: Things to Remember

1) The witness reported being "frozen" and unable to move when he first witnessed the creatures outside of his window. This again, is an obvious sign of a severe "state change".
2) The witness reported numerous nosebleeds from childhood. I cannot stress how many of my witnesses that

have endured lifelong UFO Encounters report nosebleeds from a young age!

3) The witness and family reported several instances of high strange events on the property, ie, paranormal events, strange phone calls and possible surveillance.

Case: Peeking Aliens, Part II
MUFON Case Files
Investigator: Angelia Sheer
Location: Athens, Tennessee
Date: September 2005 (Witness' best estimate; this was reported later as a historical case.)
Witness: Witness wishes to remain anonymous

Witness reports that she was in her living room finishing up some homework and then watching TV as the events of that evening began to unfold. Her modest apartment was located across from the university she was attending at the time and was not considered a rural area by any means. It is noteworthy how these events can happen anywhere as I have had sightings reported in the middle of big cities, suburban areas, and even in high-rise apartment buildings.

After she finished watching TV for the evening, the witness reported that she got up and went a short distance into her bedroom. As she came into the room, her eye was drawn to the bedroom window that was right across from her view and looking in at her were 3 alien-like creatures. At that moment, she became frozen with fear and terror. (Sound familiar?) She could not see actual sharp details of their faces but could clearly make out the outline of their bodies and heads. Two of the beings the witness estimates to be between 4 and 4-1/2 feet tall; the other was around

5 feet tall. The witness described the taller being as standing between the two shorter entities. She describes their bodies as being very, very thin. Their heads were kind of an oval/oblong shape and were supported by a very thin neck. She stressed again that she could not make out any facial details due to lighting.

She stood frozen there for what seemed a very long time, but she feels it was only a few seconds. She stresses the overwhelming feeling of being unable to breathe or move. She also experienced a horrible sense of dread like she had never felt before in her entire life. The witness had trouble verbalizing her feelings of that night and felt that there are no words to adequately express the kind of terror and sense of evil intent she felt. The witness stressed that whatever their purpose, the beings were not there for anything good!

The next thing she remembers is slamming the bedroom door and running back into the living room where she slept for the night. She did share her experience with a friend on the Internet due to her intense fear, and a copy of that narrative is included in this report. Read carefully through this conversation and notice how quickly the exchange moved from a terrifying event on to more mundane topics. This is important as it indicates a very profound state change in the witness within a very short period of time.

As with many other individuals whom I have spoken with, the witness shared that she is Christian and was praying intently for help during this event. Nothing else happened that evening, and upon questioning, it seems this sighting was a one-time event. I expressed to her how fortunate she was in that others still have ongoing encounters with beings, sometimes for the remainder of their lives. The witness describes herself as a scientist at heart, so the next morning she did explore the area in question where the beings were standing outside of her window. She found no footprints, and everything seemed undisturbed. She is sure that they were looking in at her as there are no other windows on that side of the building except maybe on the upper floors.

The witness actually sent me a drawing of her apartment and how it was set up as well as some sketches of the entities themselves. They are the classic Grays with the big eyes. I've been doing this research for 35+ years and spoken with thousands of witnesses, and it still amazes me how so many describe the same type of being. In my early years, sometimes I would test witnesses to see if I could possibly influence their descriptions of the beings. I would ask leading questions like, "They were tall with black hair, right?" Invariably, the witnesses would adamantly deny that description, stating the beings were short, skinny, and gray with big, black eyes. Many of my professional colleagues would express their concern that these individuals were delusional or experiencing a common delusion. People just don't have that kind of specific common delusion, and that explanation never held water for me. The individuals I was hearing from handled their lives just fine. They had families, managed jobs, paid their bills, and exemplified no signs of mental disorders or delusional thinking. The intrusion into their lives of these entities, UFO sightings, and other paranormal happenings were the abnormalities. This brave witness was no different! In closing our interview, I asked the witness if there was anything else she would like to share. The only other thing that she felt was important to communicate was the feeling of sometimes being "watched" at night.

Case Update: The witness contacted me a few weeks after our

initial conversation to let me know that she actually found the online conversation with her friend from the night of the sighting. When I read the notes, I was shocked to see a reference made to going out to her car and seeing the entities when she was outside. In our initial conversations, no mention was made of this. When I pointed this out, my witness again became upset as she did not remember going outside at all. Her only memory was seeing the beings through her bedroom window. I cannot say how often this happens, where memory is impaired and/or fragmented regarding these events. Many times, there is a missing time element. In this case, however, it happened to the witness so long ago, we could not determine if this was a component of her event! See conversation below:

Witness: I'm scared!

Friend: What is wrong?

Witness: There's three alien looking things in the parking lot…which is right outside my bedroom window.

Friend: OMG

Friend: Close the blinds

Witness: They are

Witness: I'm like freaking out

Witness: I saw them when I went out to my car a few minutes ago. They didn't move

Witness: but still scary

Witness: I'm seriously about to cry.

Witness: I hear people upstairs talking now

Friend: oh gosh

Friend: I am so sorry

Friend: It sucks living alone sometimes doesn't it?

Witness: Yes, it does

Witness: The best part is that while I was in my car I leaned over and accidently honked the horn.

Original Artwork by Witness "A"

At this point, both parties began talking about some very mundane, unrelated issues, which is very strange and also common in many cases. Some professionals have placed this in the category of "denial", but I feel that there is much more going on around this phenomenon. It seems, at times, that some kind of outside influence has affected the witnesses and has just directed them to "forget"! Individuals experience life-changing events, and within a few minutes, they stop talking about the sighting and then may not discuss it again for years. When I point out how strange this is to my witnesses, most have the same reaction of confusion and, many times, fear!

As time was approaching for the book to go to publishing, and we were finishing up with permission forms and last-minute edits, the witness innocently sent me this text along with her drawing:

"If they are going to study us, they should use some manners and not be peeping toms. I'm a scientist. I get it. They're curious. I know that and probably many others would be more than happy to participate in their studies if they asked properly. I would not mind!" - Witness "A"

The witness was articulate and well-educated. When we reached the point in the interview where I asked her to describe the entities, I could immediately detect stress in her voice and breathing patterns. I slowed the interview, assured her she was in complete control, and stated that we could stop at any time. She calmed herself, and we were able to continue with the interview. I stress this point as whatever my witnesses are encountering, it has caused emotional trauma to some extent. These physiologic stress responses cannot be faked and show up in a very high percentage of cases where witnesses have had close contact with this phenomenon. Also, these witness' experiences were separated by distance, age, and time frames, but notice the eerie similarities. They both described 3 entities looking in their bedroom windows, they described a very similar body type, and they both could not stress enough the terror they felt during the events. This is an example of just two cases, though this pattern of entities "peeking in the window" has repeated itself over the years with hundreds of individuals describing this scenario. Grounded, stable individuals know what they see and experience. These individuals wanted no notoriety. Most wished to stay anonymous and just wanted someone to share their stories and to be believed.

I found no evidence of hoax, embellishment, or delusional behavior. Over the years, as cases mounted, it became glaringly apparent that something was going on, and the strangeness just kept pouring in!

Author's Note: Things to Remember

1) The witness was just going to bed when she saw the beings outside her window.
2) The witness has no conscious memory of ever going outside, and we only discovered that information when she retrieved the conversation with her friend off an old computer. Indicators of fragmented memories and/or missing time.
3) Notice how fast the conversation changes from a very terrifying experience to mundane issues. This is a high indicator of a rapid state change. Have you ever been dreaming and wake up and the memory is there but in an instant, it's gone? Memory is "State Dependent"; that means that wherever the event takes place, the memory is imprinted there. To retrieve the memory, you must replicate the original state it was created in!

3. UFO SIGHTINGS... LIVES CHANGED FOREVER

"A single event can awaken within us a stranger
totally unknown to us. To live is to be slowly born."
- Antoine de Saint-Exupery

In this chapter, I would like to share some cases that are near and dear to my heart. I have worked so closely and for such long periods of time with so many of my witnesses, they and their families have become my dear friends. I have been honored to share in their incredible sightings as well as the tumultuous aftermath these events cause in people's lives. It will be hard for many of you to truly understand the emotional and psychological havoc these encounters precipitate. It's as if many have an appointment with destiny. I know this sounds dramatic, but when the overall view is examined from the inception of events to acceptance, a pattern emerges. For those who can struggle with the massive internal restructuring these sightings initiate and then finally integrate these changes in a healthy way, a radically transformed individual emerges. I have seen overwhelmed, terrified, confused people take hold of these mysterious events and, given enough time and courage, emerge on the other side with a completely different world view.

They relax...a sense of awe and hope returns to their worlds. It's as if these events take a wrecking ball to our sometimes-stagnant existence, preparing the way for a whole new place of dwelling to be built. They take up new lives; new endeavors and a sense of purpose and a new sense of adventure takes hold. Many are obsessively driven for answers and are constantly reading about, studying, and/or discussing what happened to them. Again, I really want to stress the word "obsessive". It's as if this event takes over everything in their lives. Individuals quit their jobs, relationships are affected, and the center of their world becomes completely entangled with finding answers to this mystery that has so boldly intruded into their lives.

I have had numerous witnesses copy and send me 100-page journals containing painstaking descriptions of sightings, dream events, paranormal events, and detailed drawings of encounters with ships and entities. I wish everyone could see my office and the stacks of drawings and narratives that line my desk and all available surfaces. These "chosen" are driven to paint, color, draw, and document every last detail of their mysterious journeys. It reminds me of the scene from the movie "Close Encounters" where Richard Dreyfuss becomes obsessed (there's that word again) with expressing the image that has been placed into his mind. He sculpts his mashed potatoes and then harrowingly builds a model of the implanted image by breaking out windows in his home. Individuals call me at all hours of the day and night in fear as encounters continue to happen to them. They hear strange sounds, feel things moving around in their bedrooms, and, at times, feel compelled to go outside at the strangest hours. Just the other night, a witness called me and said they were just going about their nightly routine when they were suddenly overcome with a feeling to go outside "now". They captured some type of craft that we still have not been able to identify. And…the pictures, video, and strange tales just keep pouring in.

Again, I am grateful for the courage of my witnesses who decide to come forward. I am renewed daily in my interactions with my "people": The brave ones, the tenacious ones, the ones who have been taken into that far country and have returned to tell the tale. So, let's go deeper into this mystery and take a look at some pretty "up close and personal" UFO sightings. The plot thickens, the road becomes a bit narrower, and we are drawn into this mystery, almost to the point of no return!

Case: UFO Flyby

MUFON Case Files
Investigators: Angelia Sheer, Don Williams
Location: Murfreesboro, Tennessee
Date of Event: March 31, 2015
Witness: "D"

This is one of my favorite cases involving a father and son. Both are military aircraft enthusiasts, are trained observers, and had no belief in the UFO phenomena. I have interacted on a regular basis with this witness, and he has shown himself to be both highly informed about aircraft as well as incredibly observant. He has attended many of our MUFON meetings and has been a valuable asset to myself and my team in the identification of many types of military and other aircraft. I cannot count the times that I have sent him video and/or pictures to evaluate and have been truly educated with his eye for details, patterns of flight, and general overall knowledge of aircraft specifications and dynamics. I feel honored to have been included with him and his son in his series of mysterious events.

Like so many other cases that I have investigated, the foreshadowing of the major event involved the sighting of an object at a distance but was still strange enough to catch the attention of a trained observer. I would like to share the description of this event from the witness himself. Witness "D" was kind enough to put into his own words the beginning of his life-changing series of sightings.

Sighting #1, Witness"D" in his own words:

"I, Witness 'D', will be describing the events and object, witnessed by myself and my thirteen year old son, on March 17, 2015, to the best of my abilities.

On March 17, 2015, my son and I were watching TV in our living room while looking often outside for the Northern Lights. We have an all-glass storm door and we could see everything from the door into the night sky. At around midnight or a couple of minutes after, my son looked out of the door towards the Southwestern sky, over the tree line and said, "Hey, Dad, come and see this! I don't think this is a star, it is blinking different colors and is moving left to right in a weird manner and then returning to its original spot."

I looked at the object and decided to go get my binoculars, (Bushnell Product, 16x32 100ft. at 1000 yards). What I saw with the binoculars was very puzzling and I couldn't believe my eyes. I saw a starlike object with a bright constant white/clear light. Within this bright clear light were alternating bright, like LED lights, of red, blue, and green. The lights were alternating and blinking vertically up and down. This in no way appeared to be any type of normal aircraft! I checked my watch to begin tracking how long the object was present. I noticed it was 12:08 a.m, or 00:08 military time CDT. My son and I watched the object for about 10 minutes and the more we watched, the more concerned I became. I decided to call the Rutherford Co. Sheriff's Dept. non-emergency number to report the object and to have official confirmation of the sighting.

I tried to take a picture of the object with my phone camera, but nothing was captured as the phone was not able to track or zoom in on the object. I had no other means of capturing a picture or video at that time.

While waiting on the deputy to arrive, my son and I watched the object constantly. About 20 minutes had elapsed when the object slowly and steadily began to move in a downward direction behind the tree line heading in a south/southwesterly direction. As the deputy arrived on our street, the object moved down behind the tree line, almost as if it knew they were approaching. (Author's note: It's amazing how often this happens.)

As the deputy was turning on our road, he approached from the east, and he had his clear spotlights activated on the light bar all the way across. After the deputy arrived, we explained to him what we had witnessed and the duration of the sighting. My son and I could still see the object slightly through the trees (no canopy present), so we moved to our neighbor's yard to try and get a better view. It was no use as the object was no longer in view from any vantage point. At this time, we walked the deputy back to his car, still trying to explain the unusual nature of the object. I felt the deputy did not believe us, and he then subsequently departed. (Author's note: Another example of veiled skepticism by authorities and another reason so many witnesses fail to come forward.)

After the deputy left, my son and I discussed what we had observed and were both confused and baffled by the overall event. We really could not believe our eyes! My son and I are military aircraft enthusiasts and amateur backyard astronomers. I have researched and observed most of the U.S. military aircraft in flight and in static display for over 32 years.

I, also, am a military aircraft modeler and have done extensive research over the years to ensure the authenticity of my models." - Witness "D"

I want to express here that my witness is not an alarmist and would never have made such a report unless he felt the situation was possibly threatening or completely unidentifiable. And if this event was not strange enough, it seemed to herald a close-range sighting that would forever affect my witness and his son! Only a short time later, both he and his son would experience a sighting so close as to render misidentification impossible. It is as if this phenomenon, not completely satisfied with the witness's first reaction, comes back to finish the job with an event so spectacular that the witness and his son had no choice but to radically change their opinions as to what is possible. I cannot stress enough the turmoil this witness endured after his sighting, as well as the sleepless nights, the fear, the loneliness, and the isolation these events bring. Most witnesses never ask for these events to happen, many wish they never did, and most simply want to go back to their past lives and just forget. Unfortunately, this never seems to happen, but as this mystery continues to unfold in these individuals' lives, other, greater, personal transformations began to reveal themselves over time.

Second Event
Location: Murfreesboro, Tennessee
Sighting Date: March 31, 2015
Sighting Time: 5:00-5:30 p.m.
Number of Witnesses: 2 (Witness "D" and his son)

Sighting Details

The object in question came in from the north, moving south, traveling about 80-90 mph. Witnesses report a triangular craft, approximately 300 feet wide and perhaps bigger, with a red light in the middle of the bottom of the object that slowly pulsed in intensity. The witness stressed several times how huge the object was! The craft had white lights on the corners, similar to fluorescent lights, very soft lighting that also pulsed in intensity. The lights never did go out completely. Witness estimates altitude at approximately 1,000-2000 feet. The object tracked along the witness' road across the street. Before we lost sight of the object, it had made a hard turn from a southern direction and moved off to the west before it seemed to just disappear. When the object moved parallel to the witness' point of view, it appeared so thin that it just vanished and none of the lights that were previously displayed on the top or bottom could be seen. The witness repeatedly stressed how strange it was that when the 3-dimensional object turned parallel to their vantage point, it seemed like it was almost 2-dimensional, and then it just seemed to vanish before their eyes. See recreation of the object following.

Witness "D" second sighting in his own words:

"A couple of months later, my son and I had a second sighting, and this really changed our lives forever. This sighting was a large, low-flying triangle almost directly overhead. After experiencing this sighting, I can truly state this event has driven myself and my son into extensive UFO research in the hopes of finding answers to what we both witnessed. Before the sightings, my son and I would never have thought of researching UFOs because in our minds, there was always a rational explanation. Before our sightings, we never watched programs about UFOs and would have written off such reports as hoaxes and/or other explainable events. But after what we experienced, we now know firsthand the fear, confusion, and overall astonishment others experience after such sightings.

*Late winter 2015, it was sometime near sunset, almost dark, 5-5:30 p.m.,
with a cloudy sky. My son, age of 13 at the time, was playing with his
PlayStation video game in the sunroom facing due north. His view of
the outside sky in the sunroom includes a span of due north, east, and
south from the back of our house. While playing on his video game
system, a very large and low-flying aircraft grabbed his attention directly
northwest in the sky. I was sitting in my recliner watching TV in the
living room located in the front of the house. When he saw the aircraft,
he jumped up and started to run to the front door and said, loudly and
excitedly, "Daddy, I don't know what this is, but it is large and flying
very low, it looks like it's going to crash!"*

*I jumped up, and we both went out onto the front porch. We both
observed a very large, low-flying, dark triangle with 3 large, perfectly
round lights on each corner that had a fluorescent light looking color.
They were not a true white color. The triangle craft size was
approximately 300 feet wide and probably bigger. The 3 large, round
lights at each end stayed on the duration of the sighting, but slowly
dimmed and would return to the original brightness. The 3 lights were
not very bright to begin with and did not illuminate the ground in any
significant manner. The triangle aircraft was flying below the cloud deck
very slowly at an altitude between 1,000 and 2,000 feet. As the triangle
craft slowly approached from the north, we noticed it was totally silent
with no engine noise. When the craft was nearly overhead with the one
of the outer tips just in front of us while looking directly west, the craft
slowly started to bank hard -- nearly sideways -- towards the west-
northwest direction. During the duration of the event, we were able to
see the whole bottom side of the aircraft without losing sight of it. The
key things that stood out while we were observing the craft were its size,
altitude, speed, and silence. As the craft was turning, my son noticed a
blood red colored light in the center and bottom of the triangle. He said
the lights on the corners of the craft slowly dimmed and brightened. As*

the triangle was moving away, the trailing point had 1 of the 3 round lights in the rear. It then silently disappeared! We didn't experience anything physically during the event or after.

I would love to talk to friends and family about this event, but I feel I can't because I fear they will not believe me, think that I just made it up, or worse, that something is wrong with me mentally. In some ways, I really would not blame them because before my sightings, I would have probably thought the same thing. My son experiences the same issues I do, but it's much harder on him being younger and in school. If he should tell of his experience to his classmates, there is a high probability of him being bullied and/or being made fun of. He really wants to talk about it just as I do, but we have to be careful of who we tell because of the stigma this kind of thing brings with it. I worry that due to his age, he will have to carry this burden around with him for the rest of his life. If he were to say anything to his peers, he may be known as "that kid that sees UFOs." My son and I are torn at times, sometimes wishing that we had never seen them and then simultaneously glad that we did and watch the skies in hope of seeing another one. When we saw the large triangle, both of us stood in awe and could not believe what we were seeing. We still talk about it sometimes and both look at the skies differently and pay a lot more attention to everything we see around us every day. It has dramatically changed our senses and is almost like we stay on high alert to listen for unusual sounds and scan the skies for unusual crafts. To sum things up, this experience has changed our lives forever and, in many ways, has made our lives lonelier. If it wasn't for MUFON meetings to have as an outlet, I really would not know how to handle it. In attending these meetings, we both have gained the comfort in knowing that we are not alone in seeing these craft. I just hope other witnesses come forward and will not be scared to talk about what they have seen." - Witness "D"

Triangular UFO recreation by Witness "D"

Author's Note: Things to Remember

1) Both witnesses stress how this event has affected their lives, and they have many times wished that it had never happened.
2) Witness "D" stresses the alienation he has endured due to this sighting and describes it as a great burden that he and his son must carry.
3) Even though the event weighed heavily on father and son, both were forever changed in their beliefs about their world and the possibility of life elsewhere.

Case Name: A Drive in the Park

MUFON Case Files
Investigator: Angelia Sheer
Location: Near Collinwood, Tennessee, just off the Natchez Trace
State Parkway
Date: October 11, 2013
Witness: "E"

This is another of my favorite cases! Multiple credible witnesses
are out for a drive on a colorful fall day and experience an event
that will forever change their lives. As so many other witnesses
before him, Witness "E" felt obsessed to document and share all
possible details about his sighting. I wanted to share his actual
notes that he sent me so that my audience has a feel for the work
these individuals put into their reports! Not only did he mail me
these notes and drawings, he actually went to a local hardware
store to obtain color swatches so he could accurately describe the
colors of the craft. This case is also indicative of the "high
strange" factor of many of my cases. There were two witnesses in
the truck, and although they both saw the craft, they reported
different views of what they each saw. Remember they were only
sitting a few feet apart in the front seat of the witness' vehicle.

The driver reported seeing a golden-like orb come from the back
of his truck and fly toward an egg-shaped object that appeared in
front of the truck hovering over a pasture. See witness' original
drawings and our recreations that follow. The object at one point
became partially transparent, allowing the witness to see inside
the craft. There he saw some kind of rotating column. He did not
actually see the golden orb enter the craft but feels that was where
it was heading. After just a short time, the witness describes the
object changing colors to a liquid-like silver on the top, and then it
just vanished.

The second witness in the passenger side of the truck described her sighting slightly differently. I quote, "The sky was clear blue, no clouds. I saw the sight in the sky; it was like floating very slowly. The outline was wavering. It was like a big opening, no doors, could see all the way through it. It was a blue color. The outline around it was a few inches wide. It was a bright, shiny color. Saw it about 7-8 seconds and like the snap of your fingers it just disappeared. I had no physical side effects."

Take note in the struggles many have in trying to describe something that completely falls outside of their world views. Many times, there really are no words available to accurately describe the incredible objects seen or events that transpire. Pay close attention to the details of the drawings and how much time and effort it took for my witness to record these events. And this case is indicative of 100s that I have interviewed and actually investigated. It is my experience that most individuals who are perpetuating hoaxes don't spend this much time on their reports. Matter of fact, when asked for notes, drawings, and other details, we never hear from them again!

Following are the actual notes and drawings from Witness "E". Please pay attention to all the details that went into this report. In addition, the witness actually sent me color samples from a hardware store to be as accurate as possible.

Also, I would like to make a statement about the character of these witnesses. Both were credible, honest, and humble about their sightings. When asked why he thought it happened to him, he answered, "I just happened to be in the right place at the right time." He never made any statement of being chosen or special, and he did not ask for attention or any kind of monetary gain for sharing this account. I bring this to the reader's attention as it is indicative of the character of the majority of the witnesses I deal with. Because of concerns for his girlfriend and her health, he took 3 years to come forward to report his sighting. He did not want her stressed or overly tired from too much exposure over the sighting from neighbors, investigators, or just the curious.

Eventually, as we worked together, trust was established, and I made him aware of how many others were out there like him, he agreed to actually be present at one of our MUFON meetings. There he spoke with the group about the details of the sighting and answered questions from the audience. The group loved his candor, attention to detail, and his warm sense of humor concerning his case. He has also spent hours working with me on making sure the facts of his sighting are reported accurately and continues to be a good friend and important investigator in his own right. I stress in lectures and radio shows as often as possible how our legal system is based on the testimony of credible witnesses and how this testimony is used to either convict or clear someone of a crime. It's appalling that these same witnesses' testimony concerning UFO and other high strange events is often met with ridicule and scorn. My witness felt that if he came forward and actually faced people with his story, maybe others would find the courage to come forward also. He wanted others to know that they are not alone, that we are being faced with a great mystery, and that there is a place to tell their stories in safety with other individuals who have experienced the unknown. It has been an honor to have worked with witness "E", to call him my friend, and to be included in his exploration of the mystery of the UFO phenomena.

On Friday Oct 11, 2013 (I) Witness "E" and Anon. were traveling North on the Natchez Trace Parkway, approx. 5.92 miles south of Collinwood, TN. (Per Google Straight Line Measurements.) Mesurement was take from Natchez Trace entranch in Collinwood.

Weather was post-frontal, No clouds, blue Sky.

First observation was a shiny orb traveling from right rear To Left front of vehicle. Position of vehicle was approx. 250 FT South of where Big Cypress Road (county Road 1763) crosses the Natchez Trace Parkway.

Orb was traveling at high speed and left a golden thread in It's path. Both golden thread and shiny orb disappeared in approx 1-½ Seconds.

When I Looked back onto highway, two green/grey craft were seen straight ahead at approx. (1,666 FT. distant. (Distance was determined from Google straight Line measurements. Craft elevation was even with vehicle rearview mirror. Craft were hovering side-by-side with No sound or Movement Craft appeared To Be 30-40 FT in heigth.

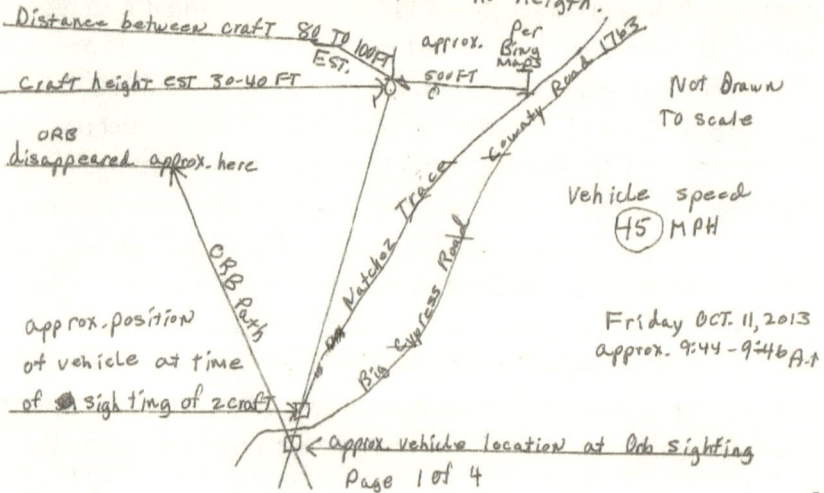

Distance between craft 80 To 100 FT EST.

approx. Per Bing Maps

County Road 1763

Craft height EST 30-40 FT

approx. 500 FT

Not Drawn To Scale

ORB disappeared approx. here

Vehicle speed (45) MPH

ORB Path

Natchez Trace

Big Cypress Road

approx. position of vehicle at time of sighting of 2 craft

Friday Oct. 11, 2013 approx. 9:44 - 9:46 A.t

← approx. vehicle location at Orb sighting

Page 1 of 4

Outline of craft can't est be described as barrel/oval shape.
Craft on left was focus of my attention. —
Lower left quadrant of craft became transparent.
Inside was observed a vertical cylinder, grey in color.
Size of cylinder est. to be 5-6FT in diameter, 12-15 FT in height.
• Direction of rotation of vertical cylinder could not be determined.
Rotational speed of cylinder est. to be that of a drill bit
at high speed in hand held drills.

Observed time est. 3-4 seconds

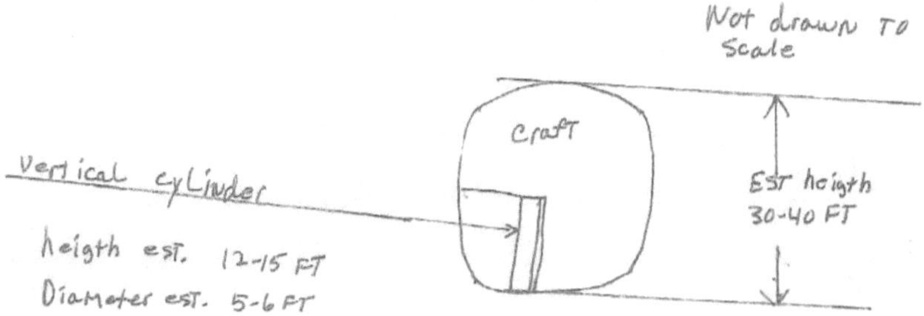

Not drawn to
Scale

Craft

Vertical cylinder

Est height
30-40 FT

height est. 12-15 FT
Diameter est. 5-6 FT

Drawing of craft
No good representation
of craft shape
Demonstration only

65

During observation of ertical cyLinder a color of
pale purple was Noted on side of craft.
A section of the purple area aligned with a "grain".
This "grain" can best be described as when a width of
hair is placed at 90° to other hair, same color, different
alignment.

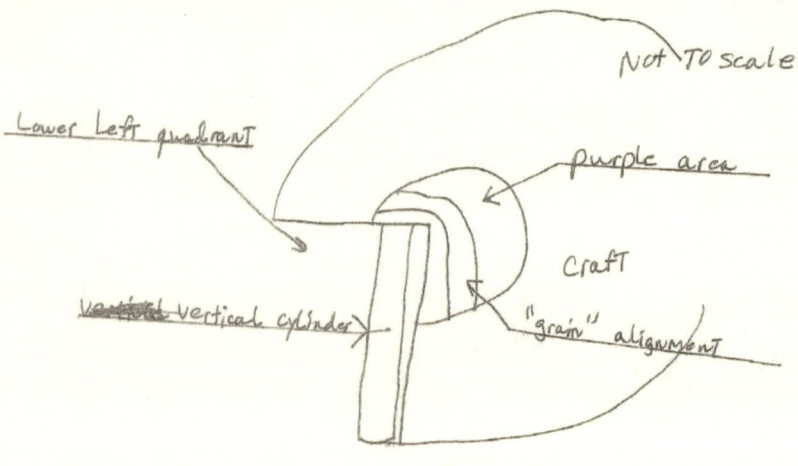

Not To scale

Lower Left quadrant

purple area

Craft

vertical cylinder

"grain" alignment

Observed time est. 1½ to 2 seconds

Drawing not good Depiction
of craft

Merely showing
purple area

At this point a flat sheet of pure silver was
Observed at the top of both craft.
  This Sheet appeared 2-3 Ft thick, both top and
bottom appeared smooth.
  Edges, while not uniform, were smooth.
  Both craft then instantly disappeared, at the same
time. There was no movment or sound noted with
either craft.

Observed time est. 2 seconds.
Observation of this event took place in clear skys.
Total event time estimated at 8 to possibly 9 seconds.
No effects of any kind were noted as a result
of this event.

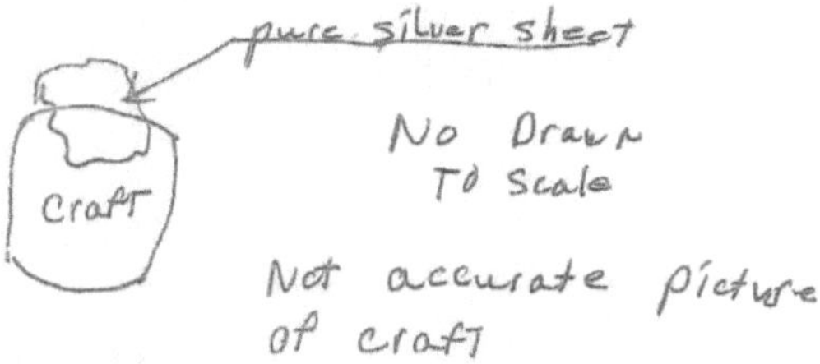

pure silver sheet

Craft

No Drawn
To Scale

Not accurate picture
of craft

All information presented is believed to be
Complete, true and accurate.
ALL information is mine and mine alone.

Angelia,

Enclosed are two purple samples that resemble the colors observed at the center of the ufo.

Below is a drawing of the white and black spots on the purple color.

. The circles were white, not perfectly round, with soft edges,

The black spots resemble coarse ground black pepper.

```
o   . o . . . o
   o .   . . .
o .   . o . o .
```

If you wish to speak to (Anon.) let me know and I can easily arrange it, at your convenience.

Color Swatches

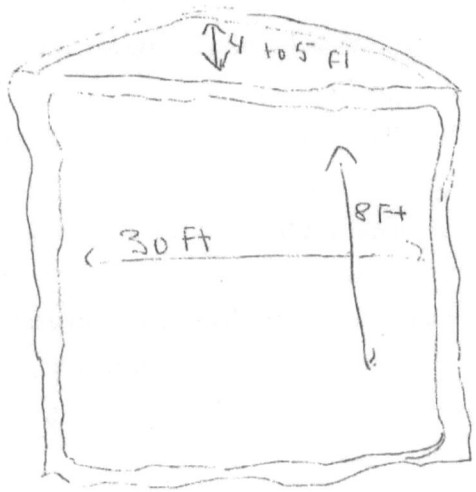

4 to 5 ft

30 ft

8 ft

The sky was clear Blue no clouds
I saw the sight in the sky, it was like floating very slow.
the out line was waving like this ~~~~ it was like a big
opening no doors could see all the way through it. It was
a blue color, The out line around it was a few inches
wide. it was bright shinny color. Saw it about 7 to 8 sec.
and like the snap of your fingers ib Just dissapered.

had no Side Effects like sickness, sight or smell

Author's Note: Pay close attention to the second witness' account
and how they describe the object in question. Their description
was very different than that of the primary witness, which is very
common in multiple witness cases. This perceptual difference
plays out again and again in my research and is an important
common denominator in the overall scheme of things.

As I shared before, these sightings have profound effects on the witness' lives. Many become obsessed with finding answers, develop anxiety disorders, sleep disorders, experience relational issues, and suffer overall life disruptions. A small percentage of my witnesses cannot integrate the radical implications of their experiences and end up disappearing from contact altogether. The others buckle down, study, deal with their fears, and, in most cases, come away with a transformed life. This is not an easy process, and it can take many years to totally integrate the radical changes these events introduce into their lives. In closing, my witness wanted to share the following:

"This sighting is one I wish I could share with everyone! It was a surreal moment in my life, to say the least. It is my hope that my coming forward will encourage others to come forward with their stories and will support them in finding the truth; that we are NOT alone! I wish to dedicate my efforts to those with open minds, curiosity, and a burning desire to know the truth." - Witness "E"

What follows is a recreation of the object Witness "E" saw that day!

Recreations by Dale Houston

This event dramatically affected the witness' life by changing his fundamental beliefs about our world and what might be possible. He continues to study and search for answers and periodically returns to the area where he had his sighting.

Case: The Journal
MUFON Case Files
Investigator: Angelia Sheer
Location: Nashville, Tennessee
Date: August 11, 2016
Time: 7:45-8:00 p.m.
Witness: John Turner

Witness Report:

"While doing yard work, I noticed a bright object in the northwest sky at about 80 degrees. I went to a neighbor's house to get additional witnesses. Two other people and I viewed the object through binoculars. I was also able to view the object through my telescope along with

another person. The object is sphere-shaped and appears to both reflect and emit light. I can't say with any confidence the size or altitude of the object, but it did appear to be rather high. As I was viewing the object through a telescope, the two other witnesses were taking turns viewing it through binoculars. We noticed that it was moving in a small circle and at the same time, a portion of the object was pulsating rapidly like a strobe light, and at the center of it was a black hole. After about five minutes, the pulsating stopped as well as the object. I had a good view of it through my telescope, crystal clear and close, it was amazing to look at. The sphere was perfectly smooth and had a bluish-white glow to it. It reminded me of the glow from an incandescent light bulb. We were talking about how strange it is and wondering what it could be. One of the neighbors viewed through my telescope and said, "I don't know what to think of that, it does not look like anything from this world that I have ever seen." After about ten minutes of the object remaining motionless, it started breaking up into smaller parts starting at the lower section of the sphere and moving up. It formed four separate "arms" made up of smaller parts that arched up and around the sphere and met into a point above it, and then it was gone. It was hard to believe what I had just seen. I think about it often. This is an unusual event on its own. But what makes it even more unusual and unique is that it's not the only time I have seen it. I have seen the object multiple times at the same location, around the same time of the evening, and the same time of the year over the past several years. It's definitely strange and one of the most amazing things that I have ever seen. I have had other impressive sightings of the object and have recorded them in a journal. I just chose one sighting that I consider to be one of the best." - John Turner

It still amazes me the detail that my witnesses sometime describe. I have been speaking with John from 2018 to the present date, and his sightings and unusual experiences continue on. I have been impressed by his knowledge of the stars, and he admits to being an amateur astronomer. When we first spoke, I thought that for sure that he had sighted the International Space Station (ISS) and had misidentified it. Then we both explored the possibility of the collapse of a high-altitude weather balloon. But, after numerous conversations, flight path, and weather balloon launch checks, I was convinced otherwise. My witness was thoughtful and articulate. He never was defensive or overbearing in his belief of what he saw, and he helped in researching every explainable possibility that we could come up with. In the end, I was convinced that the object he described in his initial report (as well as past and subsequent sightings) was unexplainable and truly landed in the camp of an "Unknown". As with many other witnesses, "JT" expressed his isolation in studying these objects and did have some concern in sharing his accounts, just as so many others do. I urged him to attend one of our local MUFON meetings in the hopes he could meet other witnesses and be able to share in a more open environment. John did attend our next meeting, and it was good to finally meet him in person. I was surprised when he brought me a copy of his most recent UFO journals. This journal was bound neatly in a folder, had a cover page with his identifying information, was complete with drawings, notes, and sketches, and contained at least 30 pages of information. I was amazed at the time, detail, and expense this witness went through to bring me this information. Seems like a lot of trouble just to make something up! Before we continue, I just want to share a sample of these journal entries and the detail they contain…

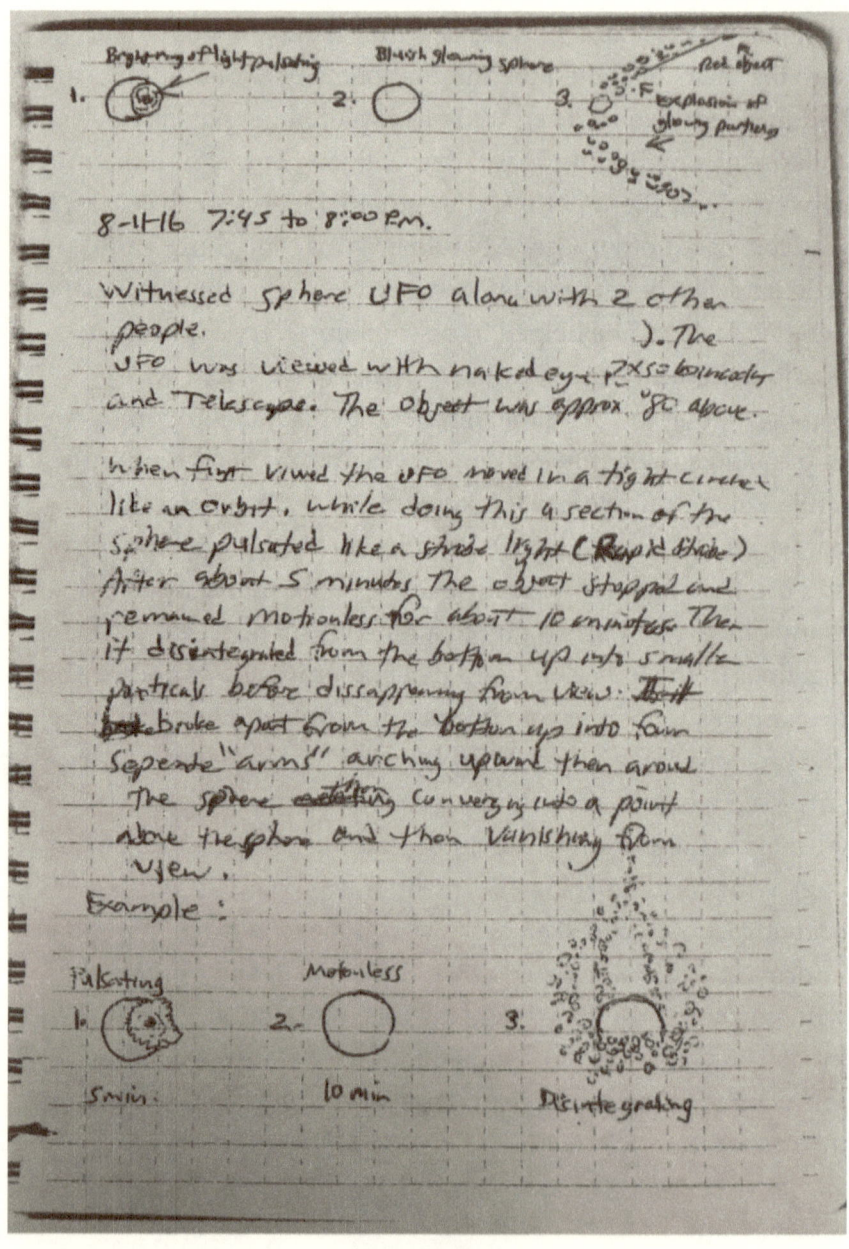

Beginning of flight pulsating   Bluish glowing sphere   Red object

1.   2.   3.   •F explosion of
                        glowing particles

8-1-16   7:45 to 8:00 P.M.

Witnessed sphere UFO along with 2 other
people.   ). The
UFO was viewed with naked eye, 7x50 binoculars
and Telescope. The object was approx. 80° above.

When first viewed the UFO moved in a tight circle
like an orbit. While doing this a section of the
sphere pulsated like a strobe light (Rapid strobe)
After about 5 minutes the object stopped and
remained motionless for about 10 minutes. Then
it disintegrated from the bottom up into smaller
particals before dissappearing from view. The it
broke apart from the bottom up into four
seperate "arms" arching upward then around
The sphere converging into a point
above the sphere and then vanishing from
view.
Example:

Pulsating   Motionless

1.   2.   3.

5 min.   10 min   Disintegrating

Original journal entry courtesy of John Turner

74

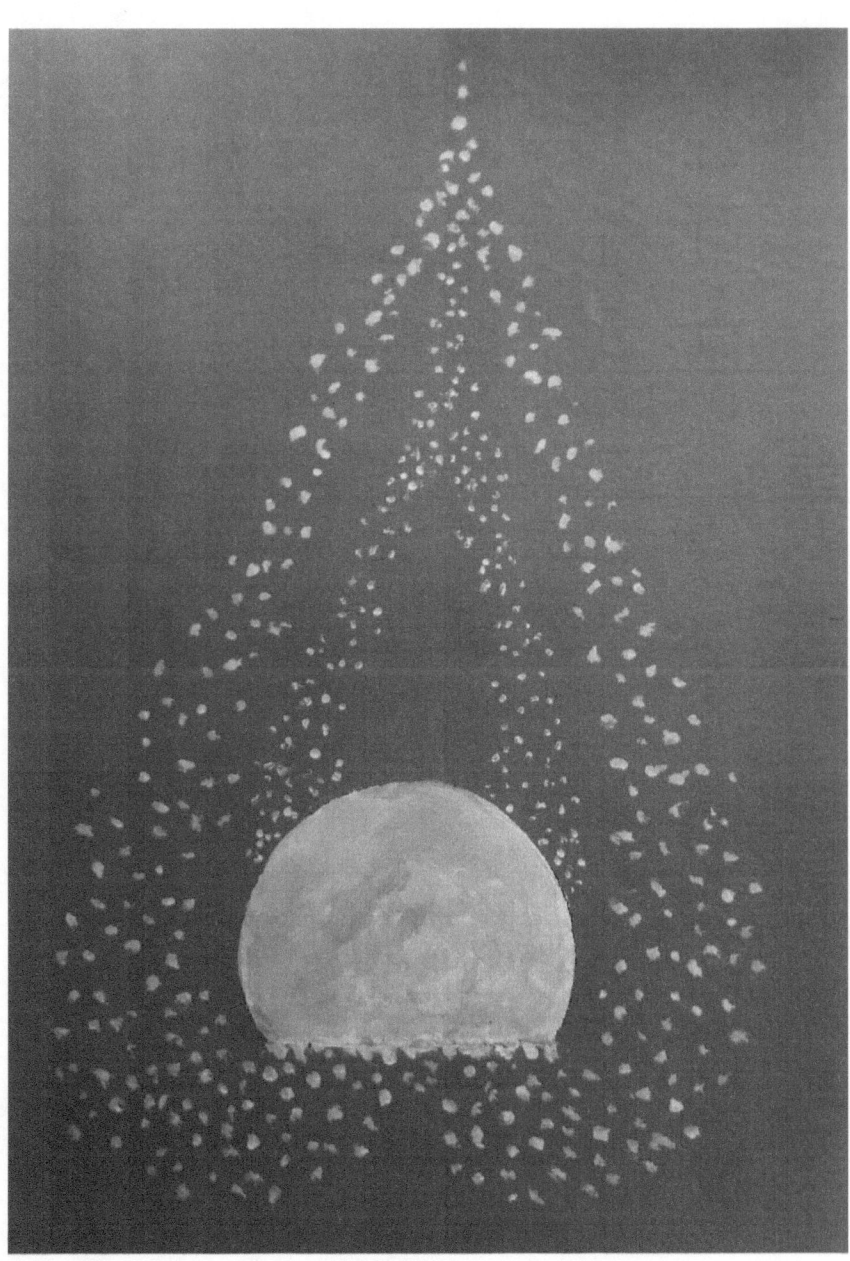

Original painting by John Turner

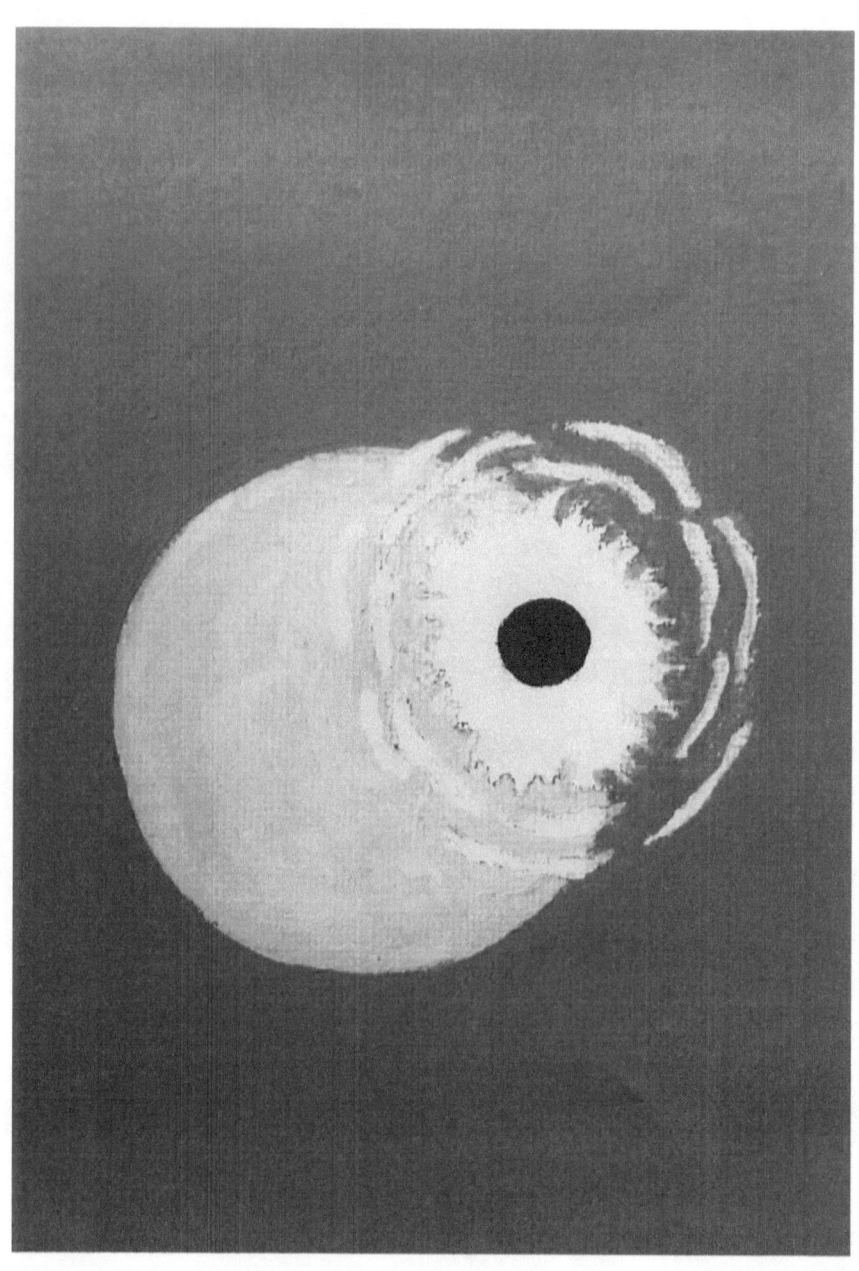

Original painting by John Turner

A painting of a UFo that my parents and I (   ) Saw in the Summer of 1980. It appeared two days in a row directly above my parents who were sitting in lawn chairs in their front yard at unknown altitude. No sound. Remained stationery for about 5 minutes. the moved slowly towards the Sun before gradually fading from view. This is what the UFo looked like through my 4inch refractor telescope. I was 14 years Old at the time. I painted the scene a few years ago from sketches that I had drawn at the time of the sighting.

Original painting by John Turner

John and I have had many discussions about his events and have spent a good deal of time deciding on which sightings, journal entries, and pictures to include. Each of the witnesses that I have chosen for this book brings with them very unique components of the UFO mystery that I wanted to share with my audience. In considering this, John wanted to not only share his UFO sightings but some examples of all the high strange phenomena that can surround these events. Again, please pay close attention to the journal entries and all the details of the reports that are recorded. I love the painting that he sent me as it not only captures an interesting daytime sighting seen by multiple witnesses but portrays the witness' obsession in documenting his experiences. So many inexplicable things that begin to transpire lead many witnesses to question their sanity. Documenting their experiences and finding others who actually share in their sightings help them stay grounded and calm their fears.

What happens so many times is an individual will have a sighting and, in a short time, other inexplicable phenomena begin to manifest. They may have multiple close-range UFO sightings, strange glowing orbs may appear around their homes or sometimes in their homes, and/or strange noises such as bumps and crashes may occur. Many times, "people" show up at their homes wearing out of date and or inappropriate clothing, strange phone calls plague them for days, and some report being followed by black SUVs. Things go missing and then turn up in places they have searched numerous times. Helicopters circle and fly over their homes multiple times of day and/or night and, when researched, no point of origin or identifying markers can be found. They can experience intense physical sensations of being paralyzed, moments of terror and ecstasy, high energy sensations, and anxiety with mild to severe symptoms of PTSD. Many times, strange entities are reported entering their bedrooms and taking them against their will. Oftentimes paranormal phenomena will become commonplace for periods of time and then inexplicably stop just as mysteriously as they started. Many witnesses report that they sometimes "just know" when something is going to happen and actually fear falling asleep. The following accounts from John's own journals are a few of these high strange events that began to follow him as time progressed.

Witness Report

"Subject: 'CLICKING SOUND' INCIDENT
Date: August 15, 2016. I can't remember the exact day but my best
estimate.
Time: Approximately 9:00 p.m.

*"I was in the garage at my mother's house pursuing one of my hobbies of
building model aircraft. The garage is a vault style located under the
living room, facing the backyard. Around 9:00 p.m., I heard an odd
sound that I remember hearing a couple of nights before. This time I
decided to investigate it. It was similar to the clicking sound one makes
when curling the tongue back and touching it to the roof of the mouth
then forcing the tongue forward. The clicks usually sounded three at a
time, and then a pause and repeat.*

*I walked out of the garage onto the driveway and stopped at the edge
where the grass begins. The sound seemed to be coming from the far
right corner of the backyard. It's not a rural area, it's in a neighborhood,
but there are quite a few tall trees in the backyard, and it's very shady in
the daytime. At night, it's difficult to see anything clearly that far back
in the yard. As I'm standing there, I hear what sounds like something
falling into one of the middle trees in the far back of the yard from above.
The next thing I hear is violent shaking of tree branches and leaves
rustling, and whatever is causing this to happen seems to be jumping
from tree to tree. At that moment, I become a little concerned because I
have never heard any animal cause that kind of loud, violent shaking
sound before, especially at night. I think about getting my 9mm pistol
and a flashlight, but I get a feeling that it would not be a good idea to do
that.*

*I decide to back up to the far side of the driveway closer to the house
because this thing is getting closer to me and still violently shaking the
trees. I'm starting to get a little scared at this point, and I feel kind of
silly doing this, but I start asking out loud, 'Who are you?' I ask this a
few times, and then I back up even more to the steps leading down to the*

basement. I feel strongly at this point that I need to go down into the basement and lock the doors. After doing that, I stand back about 10 feet from the door, staring at it, wondering what in the hell is going on. After about 20 to 30 seconds, I hear a loud, deep thumping sound that comes from the back of the house like something just slammed into the back wall of the house. I even feel a little vibration in the floor. I wait about a minute, then I slowly open the back door to take a peek outside. Thankfully everything is back to normal, no clicking sounds or shaking trees. I'm glad it's over! In the morning, I looked for any evidence of what had taken place in the backyard, but I wasn't able to find anything out of the ordinary." - John Turner, August 2016

John contacted me via email after this event, and when our schedules allowed, we finally caught up by phone. I remember him downplaying the actual events as if almost apologetic of what was happening. He felt sure something was moving in the trees but could not come up with any rational explanation of what it could be. He lives in a very suburban area with no extended wooded areas and could not reconcile how any exotic animal could be causing all the commotion. I am struck over and over how hard it is for individuals to come forward with a UFO sighting, but their discomfort increases when things just get stranger and stranger. Because of John's credibility and the hundreds of other cases that present in similar ways, I assured him that I believed him and encouraged him to just keep on documenting. I am reminded of so many other cases where witnesses reported the sighting of strange cryptids in conjunction with UFO sightings that defied all of our current scientific understanding. For some great research from an investigator that I hold the highest respect for, check out Stan Gordon's "Silent Invasion, The Pennsylvania UFO-Bigfoot Casebook". In the years 1973-1974, he and his team documented over 276 sightings of UFO/Bigfoot-like creatures with highly credible witnesses. He was so highly respected in his community; the local law enforcement agency worked closely with his team in referring sightings and had many officers share personal encounters. The activity reported was so serious and so many people were actively carrying weapons, there was a grave concern that someone would mistakenly be shot. I would have loved to be a part of that team during those years, as Stan himself and some of his team members actually experienced and documented an event that changed his ideas about UFOs and the Bigfoot phenomena forever.

Witness Report

"Subject: Strange Energy Sensations
Date: September 20, 2018
Time: Between 1:30 and 2:00 a.m.

Between 1:30 and 2:00 a.m., I awoke suddenly from a sound sleep, and I felt restless. After about 20 minutes of trying to go back to sleep, I started having what felt like waves of energy and muscle twitching all over my body. I closed my eyes, trying to ignore it, but as soon as I did that, I saw in my mind a clear image of the sphere UFO that I have been seeing since 2014. It was bright against a deep blue sky. It was as though I was looking at an HD television screen image.

It's like they were saying to me, 'It's us.' It startled me because I got an uneasy feeling that they wanted to take me. I jumped out of bed and said out loud, 'No! No!' I then fled into the living room and sat on my couch. The feeling went away soon after, and I feel asleep sitting up on the couch." - John Turner, September 2018

After this event, my witness called me the next day. This was surprising in itself since he very seldom contacted me by phone because he was always concerned with disturbing me. I was immediately aware of his high level of anxiety as he began to describe to me what happened to him the previous evening. I was immediately concerned. This man, who was never dramatic and consistently presented with a calm, deliberate manner, was visibly shaken. He shared with me that he was really scared and that he had never felt like that before. He kept apologizing over and over for calling me, and my heart went out to this kind man who was obviously in distress. I assured him that he was never bothering me and that I understood and had heard these kinds of reports from 100s of witnesses. I also assured him he was not "going crazy" and that these events are very real and affect people's lives dramatically. We talked for a while until I was sure he had calmed down a bit, and I made him promise to call me if anything else happened. Things seem to calm down over the next few weeks, and I was glad that he had no more disturbances for a bit. Later, I shared with him how many other Experiencers have shared similar stories, and we explored together some things I have found that have helped others in dealing with these extreme intrusions into one's life.

So many times, we have been conditioned to define these strange events in very predictable ways such as immediately deciding that "all" of these experiences come from external sources. Now don't get me wrong about this point. I do know that there are external agencies that interact with us, some more than others. But, as I have found in my private research, we as human beings already have some extraordinary abilities, and it seems as if those abilities are increasing. John and I discussed these points extensively as I have with many of my other witnesses. It seems as we grow, become aware of "greater realities"*, and then interact with the world at large in a more complex way, other things take notice and become interested in us! Also, there is a definitive pattern to this growth with markers and common denominators along the way. I will be discussing this at a greater length later in the book when we begin to tie all of this together but keep these ideas in mind as you continue reading through the presented cases.

Author's Note: *(I do not wish this to be interpreted in some flaky, new age way! As we grow from children into adulthood, hopefully our awareness of our world and surroundings continually increases. We are aware of our surroundings, people, and interactions in greater and greater ways. Some individuals seem to be on the fast track in this kind of development. They report all kinds of increased cognition, greater intelligence, intuitive and emotional advancements, and overall a more ordered, complex way of interacting with the world at large! Not surprising, 98% of my witnesses who experience extreme UFO/high strange encounters and endure the stress of integration present with dramatic transformations.)

To date, John's UFO sightings continue, as is the case with so many of my other long-term Experiencers. These sightings do seem to run in cycles with times of increased activity interspersed with relatively quiet times. Many of my witnesses tell me that just as soon as they least expect it, another event crops up to keep them in the game. As I was finishing up John's section of the book, he emailed me this event:

Witness Report

Subject: Pulsating UFO
Date: May 6, 2019
Time: Approximately 1:45-1:52 p.m.
Location: Nashville Tennessee
Weather conditions: Sunny/clear sky

"I was looking out the back dock door at work when I noticed a bright object stationary above the airport. The object had a medicine capsule shape to it that appeared to be pulsating. Then, a second object swooped in from the right of the first object then stopped to the right and slightly higher than the first object. After a couple of minutes, the objects started to move in an easterly direction. The lower object abruptly changed direction and started heading west, traveling a short distance then stopping. The second object then changed direction, also heading west, and accelerated to catch up to the first object. At that moment, both objects seemed to be stationary again. After a few minutes, both objects started to move again, continuing to travel in the direction of the west. During the sighting, I was the only person in the warehouse and wished I had a witness. As the two objects got closer, I heard what sounded like a box being placed on the floor. I thought it was the woman who works in the office. I called her name but didn't get a response. The objects started to pass almost directly above me. I'm guessing they were at about 500 feet. I then could see that the shape of the objects was similar to a boomerang and rotating slowly in a counterclockwise direction. I lost sight of the objects as they passed over the building. Interestingly, I saw the same type objects at my mom's house on Easter just a short time before. They passed overhead from west to east." - John Turner, May 2019

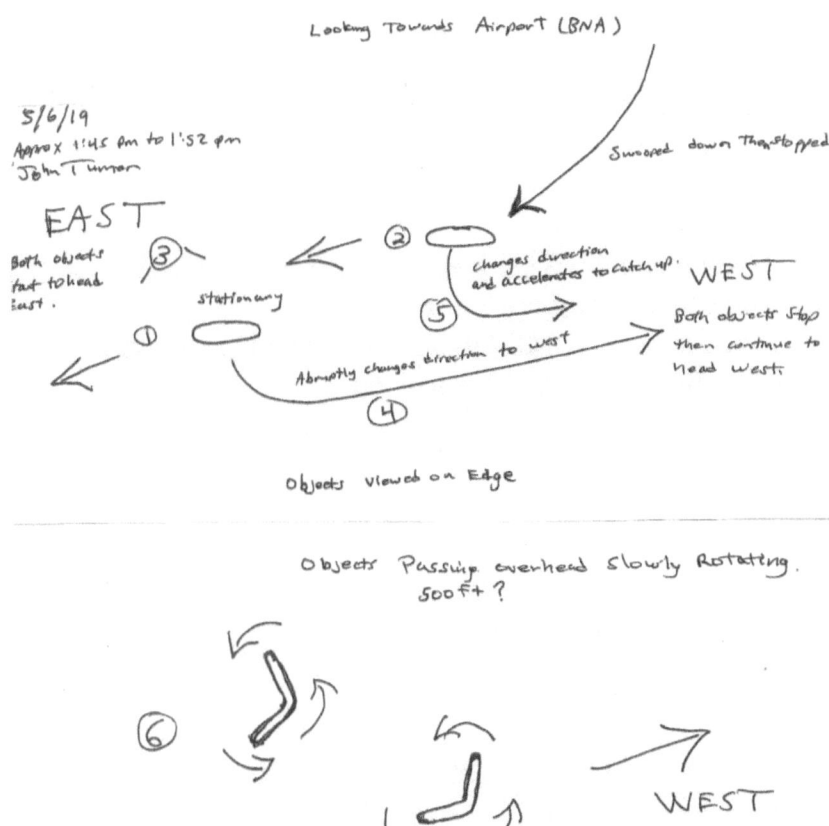

Looking Towards Airport (BNA)

5/6/19
Approx 1:45 pm to 1:52 pm
John Turner

EAST

Both objects
fast to head
east.

③

Swooped down Then stopped

②

Changes direction
and accelerates to catch up.

WEST

Stationary

⑤

Both objects stop
then continue to
head West.

①

Abruptly changes direction to West

④

Objects viewed on Edge

Objects Passing overhead Slowly Rotating.
500 ft ?

⑥

WEST

Same type objects I saw on 4/21/19 at my Mom's house

Drawings by John Turner

85

Author's Note: Things to Remember

1) The witness reports sensitivity from a young age and continues to experience UFO and high strange events.
2) *Energetic Phenomena* were reported during one experience. This is highly indicative of highly charged state changes and many times are precursors to major psychological and/or spiritual transformations.
3) The Witness' UFO sightings and other high strange events continue to be reported to date.

4. MISSING TIME

"If you bring forth what is within you, what you bring forth will save you. If you do not bring forth what is within you, what you do not bring forth will destroy you!" — *The Gospel of Thomas*

Case: On the Railroad Tracks
MUFON Case Files
Investigator: Angelia Sheer
Location: Chattanooga, Tennessee
Date: January 8, 1969
Witness: Suzie Tu (Assumed Name)

<u>Witness Report</u>

"On this particular night, my mom, myself, and two others were traveling to a basketball tournament across town. We left at 7:30 p.m. The trip should have lasted 15 to 20 minutes, tops, but we arrived at 10:45 p.m. I first noticed the craft and thought it was a large harvest moon, but I noticed it was traveling the same speed and distance we were. I told everyone to look!

The next thing we knew, we were parked on the railroad tracks (car not running) with the craft hovering right in front of the car! I was excited, nervous, confused, shaking, and wanted a photo. The two in the back seat were screaming to get the camera. I retrieved the camera from the glove compartment, and after a couple of shots realized there was no film. The craft, while stationary in front of us, was glowing and changing from blue with an orange color around it to an orange color with a blue ring around it. When it changed colors, there would be several large puffs of smoke, then all at once, the smoke disappeared, then the craft changed colors.

At this point, a train was approaching our car on the tracks and started blowing its whistle, everyone was terrified, and we started screaming for Mom to start the car. The first couple of times she tried, it did not start, and she told us if it didn't this time, we should all get out and run! Thank God the car finally started, and Mom drove as fast as she could off the tracks and on toward the gym. Mom let us out at the tournament at about 10:45. When we first arrived, I clearly remember my friend untying her shoes on the bleachers as we walked in. I was confused and asked her what was going on? She told us the tournament was already over and everyone was leaving. I said, "Quit playing with us!" I guess because of the disbelief we expressed, she told us to look at the clock. Upon checking, we were shocked to see that it was already 10:45 p.m. At this point, I think we were all still in shock as we still could not really understand how so much time had elapsed since we left home. Several people wanted to know why we arrived so late, and I told them what had happened. I was so made fun of that I just stopped talking about it, and we went to call Mom to come back and pick us up.

On the way home, Mom told us that she had called the airport earlier and spoken with the air traffic controller and asked if the moon was visible anywhere in our area that night, and he had told her, "No!" She then went on to describe what had happened, and then as she is speaking to the first controller, she overheard another person burst in wanting to know what the "hell" he had just seen? The gentleman on the line told him he had no idea but had been receiving call after call about it that evening. He told my mom, still on the phone, "Lady, I have no clue as to what you saw."

When Mom came and picked us up, we all talked excitedly about it all the way home. The next day, a short paragraph in our local newspaper stated hundreds of people saw a strange object in the sky. They stated it was a weather balloon according to the airport and weather service. We all laughed and then got mad that the government would think we are that stupid. My mom is now deceased, and the other two people that were in the car are so afraid of ridicule they deny it ever happened.

Years later, I was watching a program on TV program that actually caused me to experience some kind of flashback episode. I remember being placed back in the car and the distinct sound of the car door as someone opened it, me being lifted like a feather, the door closing with a muffled sound. I remember it was bright like the day and I could see everything so clearly. I wondered in reflection if there was a streetlight nearby or maybe the light was coming from the object? Mom was being placed back in the car at the same time as me, and then the two in the back seat. I was told you're okay, close your eyes, and don't let them see you're awake. The one in charge came and asked, "Did any of them wake up?" The one placing me back in the car said, "No." To this day, I believe, through my tears, whoever he was saved my life. I waited a long time to make certain they were gone. It was like I woke up before I was supposed to.

As a child I remember "Johnny", the name I gave a being that interacted with me throughout my childhood. (Many times children that have had ongoing interactions from a young age, name the beings they come to know). As an adult, I came to understand that "Johnny looked like a typical "Gray". He would come periodically and we would both fly over the hills where we live. We played with others and had fun, and then bad things happened. I have not fully explained to anyone the terror of being restrained, held down, not being in control, etc., and this has affected me. I have major panic attacks, anxiety issues, and health issues. Growing up, I experienced paranormal events on a pretty regular basis. My mom was clairvoyant about family matters and was very accurate at seeing events that may affect us. Many times, Mom would not allow us to go to school. One time, a sibling disobeyed and laughed at Mom and told her she was a silly old woman. Mom received a call at lunchtime that my sibling was playing on the steps at school and had broken three ribs and his arm. After that, none of us, or any our friends, would ever disobey her when she told us not to do something. We had other strange events that were somewhat commonplace, like episodes of dishes flying off shelves after being dried and neatly stacked. This happened on numerous occasions with my mother scolding me and telling me to "stop showing off". As a child, this was completely baffling as I had no idea that I could be doing such a thing, and I still wonder about these episodes. (Readers take note!)

All of my life, I have wanted to share my story, but I have never come forward because of the ridicule I would have experienced from my family and from the community. I want to do this now because of worsening health, and I want to make sure my experience is known. My prayer is that people of Earth will wake up to the facts about our visitors. Some have our best interest at heart, and others are evil and want to use us for their own selfish, ungodly reasons. We need to pray every day for our leaders to have wisdom when dealing with all of them. I know there is God, the Father, and His Son, Jesus Christ. I know He created all the worlds. Hebrews 1:2 - Hebrews 11:3 By faith the worlds were created. As a believer, I am His child adopted into His family. There is no doubt in my being about that!" - Suzie Tu

Witness Phone Interview: February 27, 2019

When I called Suzie, she was at once very grateful to finally have someone to share her story with. It's very hard for the general public to understand the burden of carrying these events alone, sometimes over 40-50 years. The witness was honest, emotionally open, and credible in the telling of her event. Over the next few months of my investigation, she was completely cooperative and helped immensely in tracking down sources of information such as names of roads, nearby airports, and local newspapers. She worked diligently in creating drawings of the craft, taking pictures of the area where the event transpired and searching the local library for articles that may have been written about the event. Even with her poor health, upcoming surgeries, and doctor visits, Suzie never let me down in her help with the investigation!

The witness began her story by describing her mom driving her and two other passengers to the gym in their 1966 Plymouth Fury III. The witness, 16 years old at the time, was riding in the front passenger seat, and at one point, she started noticing a large, yellow-orange object with a blue outline around it below the tops of the trees. The object was round, like the moon in shape, and was following the car low in the sky as it was visible in its movement through the trees. At first, she thought it might be the moon, so she asked her mom if it was a harvest moon, and her mom said no, that she thought it was too late for that. She told her mom that the object was following them, and she was sure her mom looked over to see the object. At this point, the witness and the passengers evidently lost consciousness because the next thing she describes is waking up with the car stalled on the railroad tracks and the object is hovering directly in front of the car. She has no conscious memory of how they came to be on the railroad tracks or of the object moving into this position. (Remember Jack's experience on the railroad!)

Realizing they are stranded on the tracks and seeing the object hovering directly in front of the car, the witness started

91

desperately shaking her mom who had not yet fully awakened. At this point, the two passengers in the back seat woke up next and began yelling for her to take a picture. She grabbed the camera out of the glove box and snapped several pictures but then realized there was no film in the camera. At this point, chaos took over inside the car as everyone simultaneously realized that a train was approaching their position rapidly. The witness' mom was desperately trying to start the car and finally told everyone if she couldn't get it started at the next try, everyone needed to get out and run.

In the meantime, Suzie was watching the object as it hovered. She described the object as changing from blue with an orange ring to orange with a blue ring, and every time it would change color patterns, she could see the object emit what seemed to be puffs of smoke or fog. Also during this time, the witness noticed some people standing in the parking lot of a nearby store. She describes the people as looking up and pointing as if seeing the object, but they also seemed to be unmoving and frozen into place. I have had numerous other reports that describe these bizarre circumstances. For example, one witness, while driving along a very busy highway, noticed a huge triangular object just hovering over the road. They decided to pull over to watch the object and observed one other car pulling in behind them, evidently to also get a closer look. What astonished both witnesses was the fact that no one else on this busy highway seemed to even notice this huge object just hovering over both lanes of the road! It was if no one else could see this huge craft just hanging there in the sky, and everyone just kept on driving, seemingly oblivious to the events at hand. Here is a sketch made by the witness of the position of the car on the railroad tracks and of the store and people standing outside looking up and seemingly frozen in place!

Original drawing by Suzie Tu

Finally, the car did start, and they sped off under the object to get off the tracks, barely missing the oncoming train! When they finally arrived at the school, it was 10:45 p.m. Remember earlier, the witness stated they had departed their home at approximately

7:30 p.m., and it should have only taken them approximately 20 minutes to arrive at their destination. If this is correct, then the witness and passengers experienced approximately 3 hours of missing time! When they arrived home, the witness' mother called the airport to report the object they had seen. As mentioned earlier in the witness' actual report, she was told by airport officials that they had received numerous reports of an unidentified object that evening. Interestingly, while speaking with the airport official, she overheard the controller talking with someone else who said, "What the hell did we see?", again validating that something truly unusual did transpire.

Here is the drawing of the object as the witness first observed it through the trees.

Original drawing by Suzie Tu

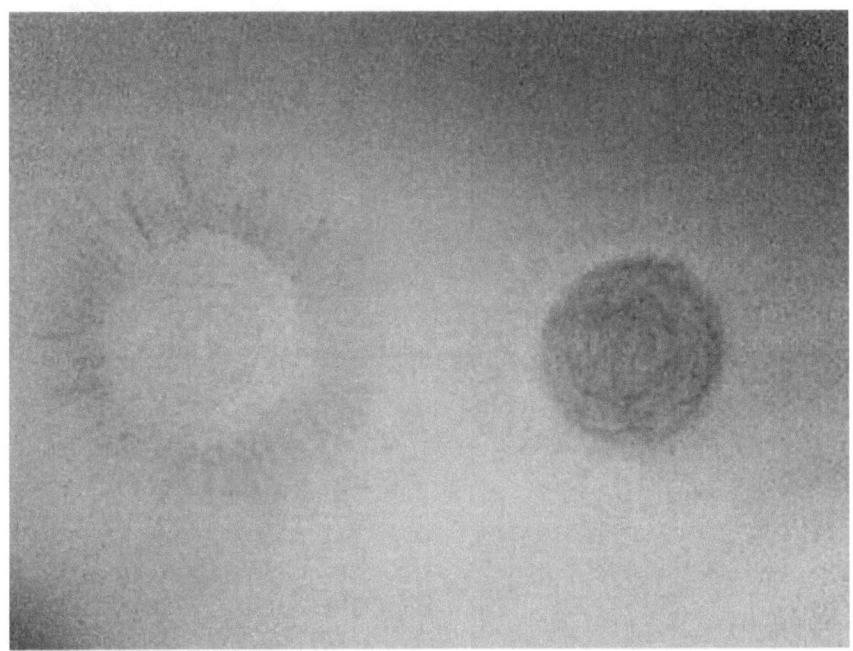

Original drawing by Suzie Tu

<u>Author's Note</u>: This is how the object appeared when it was stationary in front of the car. There was only one object, but the witness wanted to document how the object changed colors. Notice that the object changed from blue with an orange/yellow outer ring to orange/yellow with a blue outer ring. When the object was first spotted, it looked like a big orange/yellow sphere that was internally illuminated. The witness said at first it might have been mistaken for a full moon, but later after moving out of the trees, that was completely ruled out. The witness reported that her mom had called the airport after she had returned home from dropping the girls off at the gym that night to report the object to the controller on duty. She asked if there was a full moon that night as she thought they might have just misidentified the object, but she was told by the controller on duty that there was no full moon!

<u>Author's Note:</u>

The witness wanted to convey seeing people standing in the parking lot of an old country store, seemingly frozen in place. Other cases have described seeing huge, low-flying craft while experiencing unusual fogs that enveloped the car, time distortions, frozen people (as indicated above), and feelings of some kind of surreal event that surrounded them for a period of time. One witness said she was driving on the interstate, and there was a huge craft just hanging over the road for all to see, but everyone just kept driving by like they saw nothing! These are important markers in case investigations, and we will be discussing the implications of these bizarre but common reports in later chapters and how they play into the overall scheme of the UFO phenomena.

Photo Courtesy of Suzie Tu

In concluding my interview with Suzie, I continued my line of questioning to include the following inquiries. This protocol is followed with all of my witness interviews.

1) Did you experience any physical effects after the event such as headache, nausea, vomiting, eye irritations, hair loss, etc.?
2) Did you experience any lingering illness, unusual surgical issues, or other medical problems that you feel may be related to your UFO experience?
3) Did you experience any unusual dreams?
4) Did you experience any paranormal events preceding your sighting or after?

Remember my definition of "Paranormal" is any event that is not defined by our current scientific paradigms? I feel strongly that any alien civilization that possesses technology significantly more advanced than our own could appear as "magic" or as some kind of supernatural event to our limited understanding of science. Here we must also start to question our abilities as a species and (evolving perceptual abilities) as a race, latent and otherwise. I will expand on this in later chapters.

Further questioning revealed a history of chronic health issues, strange surgical experiences, and the onset of chronic nightmares. The witness shared the following:

"I remember a time I was to have surgery on my nose for a deviated septum. After the procedure was over, the doctor reported that surgery went well and trimming of bone and cartilage was not necessary. This was really strange as this is the point of the whole surgery. In the end, my breathing was greatly improved, so I believe something was removed."

"Anytime I have been in the hospital or away from home, I have to cover up the whole truth. If anyone came in my room without knocking and calling my name from the doorway across the room, I would scream bloody murder and wake everyone on the unit. The doctor talked to me about this, and I could not tell him that I thought it was "Johnny" coming for me during my sleep. They would have thought I was insane and treated me accordingly. Hiding all of this was so hard! I had to tell them it was only the stalker I had that I was afraid of. Sleep deprivation is something I deal with daily also. I would imagine other people deal with sleep deprivation, too."

"I have experienced chronic reproductive issues my whole life."

"After the incident on the railroad tracks, I remember all of us in the car having what appeared to be a slight sunburn on our faces."

Again, I cannot state how many cases I have received that included similar reports of sunburned faces, strange surgical issues, chronic health issues with a high incidence of reproductive problems, and the onset of chronic nightmares. In another case I was involved with, a witness shared that she had a strange object under her skin on the back of her neck. During an unrelated event, she had back surgery and the doctor removed the object from her neck without her consent. When questioned, no answers were ever forthcoming and of course the object was never retrieved.

Upon further joint research, Suzie and I learned that there were several newspaper articles written about a local UFO sighting during the time period in question. With a little detective work and the assistance of the staff at the Chattanooga Public Library, we were able to obtain said articles. The article was dated January 8, 1969 and was actually a Tuesday. Suzie states that's when the girls had their games in junior high school. This is a huge corroboration of Suzie's sighting and sequence of events. In closing, I want to thank Suzie Tu again for her bravery in sharing her story, her hard work in gathering pictures and drawings, and her dedication to making the truth known. My witnesses are so driven for their stories to be heard, Suzie had me record a verbal permission for her story to be printed in this book as she was having surgery and feared that if she died before her written permission form was returned, her story would not be printed. I am continually honored to work with such amazing, generous, and honorable individuals. Per my witness, Luke 8:17 says, "For nothing is concealed that won't be revealed and nothing hidden that won't be made known and brought to light."

Calls Swamp Local UFO Organization

NFP 1-8-69

The head of the local UFO group was swamped with telephone calls from Chattanooga residents Tuesday night, reporting a "large, orange-colored glow, larger than the moon," in the southern sky about 6:45 p.m.

David Kammer, chairman of the Tennessee Subcommittee of the National Investigations Committee on Aerial Phenomena (NICAP), said his phone rang "continuously for a couple of hours" by local residents reporting the bright object.

Mr. Kammer said the object, which was visible for several minutes, seemed to turn into a "white smoke cloud," and was probably "one of the special high-atmosphere chemical tests being conducted by NSA at a Florida missile base."

Similar t e s t s, where chemicals are shot into the upper atmosphere and released, causing illumination, were performed several months ago at Wallace Island, W.Va., and were also visible from the Chattanooga area.

Articles provided by the Chattanooga Public Library circa January 8, 1969, Chattanooga Times

Author's Note: Things to Remember

1) The witness reports UFO sightings and high strange events from childhood.
2) Indicators of state changes were apparent in missing time accounts as well as memory fragmentation. This is apparent also by reporting that others seemed frozen in place which is a possible description of time anomalies.
3) There was a family history of Paranormal Events.

Case: Strange Voices

MUFON Case Files
Investigator: Angelia Sheer
Location: East Tennessee
Date: August 8, 2016, 10:30 p.m.
Witness: Anonymous (Identifiers have been changed to protect the witness.)

Witness Report: Phone Interview February 5, 2017

On August 8, 2016, my witness and a friend were sitting out in his truck on the family farm listening to the radio and just enjoying the warm summer night. At one point, both young men noticed an increased static coming from the radio and then watched a strange triangular craft come into view. Both witnesses were stunned, could not believe what they were seeing, and got out of the truck to take a better look. They saw what appeared to by a huge triangular craft hovering over the treetops. As they were watching, the ship approached slowly until it was almost directly overhead. It would periodically shine a beam of light down on them every time they would speak with each other. Witness "A" kept stressing the quality of the light that beamed down from the bottom center of the craft, and he struggled with words in describing this "light". It was somehow unearthly and mesmerizing in its appearance. He remembered the bright blue colors of his friend's flannel shirt lighting up in the beam of light. He reported that their minds went blank and they could not believe what they were seeing. During the sighting, he felt moments of euphoria mixed with terror and the suffocating feeling of being watched. He noticed that there was a static charge in the air, and the hairs on his arm were standing up. There was also a strange odor like burnt copper wires mixed with a hint of cinnamon. During the entire event that he estimates was about 10-15 minutes, there was a very low droning sound that permeated the entire area.

The witness described the craft as triangular and grayish in color with bright white halogen-type lights. He estimated the craft to be at first about 200 yards away with an estimated altitude of 300-400 feet. As they both watched, the object slowly made its way directly over the witnesses' heads. After the object moved off, the witness and his friend talked about the sighting for a bit, but his friend was so "freaked out" that he left shortly thereafter. It was at this point the witness also shared with me that he and his family were farmers and that for many years around that mountain area, strange lights had been seen. He also reported that cattle had been found mutilated with eyes missing, anal regions carved out, and with no blood found at the sites. Many of the farmers said it was just lightning strikes, but he assured me he had been raised on a farm and knew for sure these deaths were not caused by lightning. Cattle mutilations have been associated with UFO sightings for a long time. Linda Moulton Howe did exhaustive research on these phenomena in her work, "*A Strange Harvest*". I found this to be an extraordinary work and recommend it to all serious researchers.

Later that night, other strange things began to happen. After the witness went to bed, he fell into a fitful sleep. At times, he felt that he couldn't move, and at other times, he felt a compulsion to go outside, like something was calling to him. There was this strange chatter in his head that was hard for him to describe to me. The best he could come up with was like listening to a distant CB radio conversation but in non-human voices. He remembers the room filling up with white light and then being in a strange room that he didn't recognize. At this point, he tried to call out but no one heard, and he was left alone to deal with what was happening to him! His fear increased to panic as he saw a gray being coming toward him, he was struggling to move, to escape, but he was paralyzed, and then he just blacked out.

The witness continued to struggle to describe to me the events of that evening, and I could hear his fear and anxiety as he tried to make sense of what happened to him that night. He shared that he felt violated somehow, and he broke down and actually cried on the phone. He kept going over and over this part of the story again and again, as if to put together the sequence of events in some kind of rational way. But every time he would reach a certain point in the narrative, it was like he would just hit a wall, almost like something was preventing him from remembering. At this point, my witness became so distraught, I had him stop and spent some time calming him down and assuring him he did not have to remember everything at once. I assured him that there was plenty of time for the story to unfold and that I would be available to him any time, night or day, if he needed me. I cannot stress enough the importance of the relationships that I form with my witnesses. In many instances, they have no one to share their stories with, are completely anxiety-ridden, and become obsessed with the event. Having someone they trust to talk with and who actually believes them provides, in most instances, a degree of relief and a calming effect.

In the days following the sighting, the witness reported extreme eye irritation, stating that "his eyes watered for days". He also reported nausea for the first few days and then generalized pain and discomfort throughout his whole body for weeks after the sighting. Interestingly, a family member asked him the next day where he was during the night. He shared with me that this almost pushed him over the edge because it indicated that he really WAS out of the bedroom at some point during the night, and this wasn't just a bad dream. Since the encounter, he reports he doesn't sleep well and that the events of that night weigh on him greatly. At the time of the interview, I asked him if his friend would talk with me, but he shared that the other young man refused to even talk with him about it and wanted nothing more to do with the whole event. He also wanted me to know that he struggled for a long time to come forward as he felt that in some way he was "not supposed to talk about it" to anyone. You don't know the times I have heard that from witnesses as if someone or something has compelled them to keep quiet about the whole affair. He also shared that he had developed a high anxiety for driving on dark roads at night, another common issue among Experiencers. He did ask me about hypnosis and whether I thought this may help him. We had several conversations about the possibility of finding a trained hypnotherapist for him and the pros and cons for this process. I explained that hypnosis performed by a trained professional was an excellent tool in retrieving lost memory, but the witness must be prepared for what may be uncovered and have the resources to deal with it. Many move forward and do well with what they uncover. Others decide they don't want to know and refuse to talk about the events ever again. Unfortunately, many of those that choose not to face what has happened to them continue to suffer with varying degrees of anxiety that affect the remainder of their lives.

As the days and weeks went by after our initial conversation, I tried to reach out and speak with my witness on a regular basis. He reported one other UFO sighting of a similar triangular object while with a group of his friends, and this seemed to only increase his anxiety and internal turmoil. I was able to reach him a couple more times, but after that, emails dwindled over time, and to date, there has been no more contact! I fear that the burden of his sighting may have been too much for this gentle soul to suffer, and I worry that he is out there struggling with his fear and anxiety all alone. I do hope that one day he may change his mind and reach out for help in dealing with the emotional trauma of his UFO sighting!

Authors Note: Things to Remember

1) Both witnesses experienced a low humming sound in association with the object as well as fear, euphoria, and paralysis, which are indicative of immediate and dramatic state changes. Things appeared to slow down and seemed somewhat surreal in nature. The witness commented on the colors of his friend's shirt and how vibrant they became.

2) The impact of the sighting on this witness was extreme, and it became apparent as time went on that he was having trouble dealing with what happened.
3) The witness appeared unable to integrate the dramatic changes that were flooding his being and eventually cut all forms of contact.

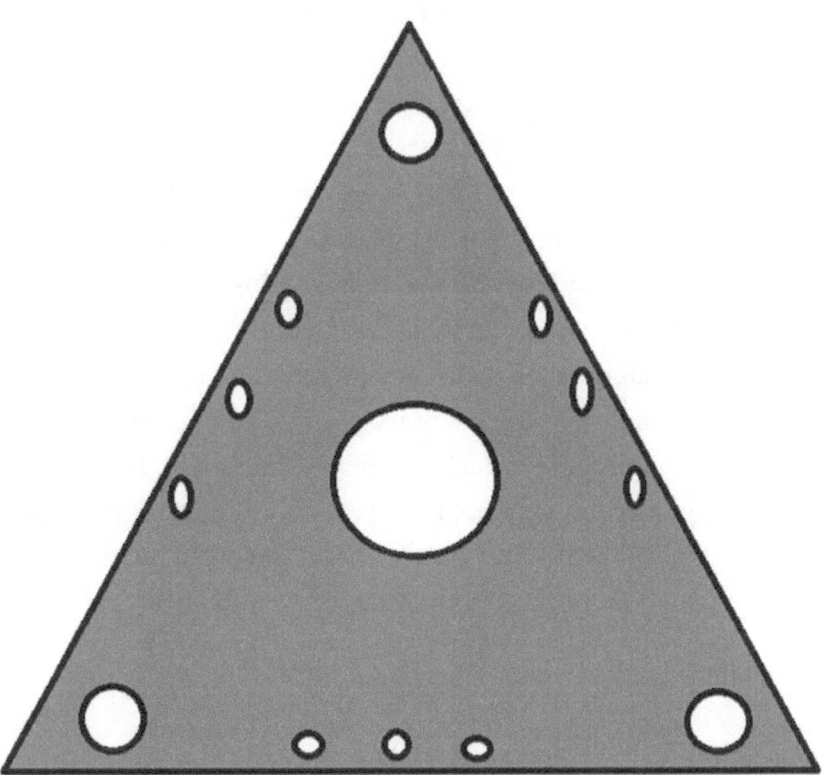

Drawing by Angelia Sheer, a reproduction from Witness description

Case: The Strange Case of Witness "J"

MUFON Case Files

Investigators: Angelia Sheer and Tennessee STAR Team

Date: April 19, 2016 into the early hours of April 20, 2016
Location: Undisclosed location in Northern Tennessee
Date: April 19, 2016 into the early hours of April 20, 2016
Witness: "J"

I remember so clearly the morning that I settled into my office to check current UFO case reports as part of my normal duties as MUFON Chief Field Investigator for Tennessee. All cases reported across the state of Tennessee to MUFON are sent to me promptly so in the event of close-range sightings, etc., we have a chance to deploy a team to investigate. When I opened my new reports, I was stunned to find one of those 1-in-100 cases. It read like the X-Files in its complexity and hinted at the possibility of some pretty incredible evidence. I was excited to call the number provided immediately as sometimes witnesses experience what I call "reporting remorse", much like buyer's remorse. They will summon up the courage to make the report but then regret doing so and will never return our calls for an interview. I hoped and prayed that would not be the situation surrounding this case.

Following is the basic report submitted to MUFON:

Paraphrase of Actual MUFON Report

"April 19, 2016 into the early hours of April 20, 2016

I was working in the shop separated from our home of about 100 feet. I had begun working about 3:00 p.m. that day until almost midnight when, for unknown reasons, I was drawn to the side door of the workshop. (Cell phone was recording from a previous video I had made in the shop area unknown to me. At this point, the phone was set down on a cluttered worktable along with my firearm.) I felt as if I was in a fog. I then woke up on my back on the driveway to my house absent my cell phone and firearm sometime later with my keys separated into both hands and the neck of my jacket pulled down to my elbows, limiting my movement. Seeing the stars and the outside garage light, I knew where I was and was alarmed yet could not understand how I got there and why I was on my backside. I used my keys to get in the house and noticed light coming from my elderly mother's room. When I entered her room, I was surprised as she was sitting up in bed with her eyes open, but I could tell she was in some kind of groggy state with my pit bull in bed with her. I immediately asked her what time it was, and she replied 3:20 a.m., and she asked where my phone was (I didn't know) and then asked where my sidearm was (again, I didn't know). At this point, I think I freaked out a little and threw my keys from a loss of words and then began to tell her, "It's here, it got me, it's evil, and it did something to me. I've lost three hours of time! After regaining some composure, we both went back outside to the garage where she covered me with her sidearm from the garage door so that I could retrieve the cell phone and my sidearm. Barn lights were off, door was unlocked, firearm and cell phone recovered yet battery was depleted. Once back inside, we took pictures of my back, my head, and my arm because it hurt to the bone when touching the skin. I had multiple red dots almost the size of a penny on my upper scalp and strange markings on my back. My mother also had strange red dots on her body. The next day, I charged my phone and noticed I had 2 hours and 29 minutes of blacked-out video with the audio present. In the audio, you can hear what sounds like language, some of it sounding human and some not. You can also hear me moaning and verbally fighting with someone. You can also clearly hear my dog deeply barking from inside my mother's room, yet she does not remember any of it and cannot believe she would have slept through all of that noise. I, myself, am a military-trained combat vet as well as the rest of the males I descend from. I have 3 hours of missing time and a 2.5-hour recording of some of what happened to me. Someone needs to contact me from your organization with the cell phone number provided. I will submit the audio/video and photos to help me get more answers as something was clearly done to me without my consent. Thank you, Witness 'J'"

Needless to say, I was anxious to talk with this witness immediately as these types of cases that may possess actual physical evidence are rare. I reached out immediately to the cell number provided and left a message for the witness to call me. I waited for 24 hours, and there was no reply. I was worried that the witness may "bolt" and I may never have the opportunity to speak with him. Eventually we did connect, and Witness "J's" apprehension in speaking with me was readily apparent, as is very common in many of these events. Like I do with all of my phone interviews, I spent the first part of the conversation just getting to know a little about each other, telling the witness a little about me and my history into the UFO phenomena, and explaining "the rules of engagement" during our course of interaction. I explain to the witness that I will ask many questions and this in no way is meant to infer non-belief but to weed out any mundane explanation of the events at hand. I also explain that they have the right to refuse answering any question that may make them uncomfortable and can terminate the conversation or investigation at any time. Over the years, I have created an investigation protocol that seems most beneficial not only for the investigative process but ensures the mental, emotional, and physical protection of the witness.

After the initial moments of the introductory phase of the phone interview, Witness "J" and I settled into what I felt was a comfortable conversation about the events that had just recently transpired. I found the witness to be a bit guarded at first, but then as we got to know each other he was open and animated in the telling of all that had transpired. As in many cases, he seemed somewhat relieved to be sharing his story with someone who was genuinely interested in what happened to him. After probably a two-hour conversation, we both agreed that a field investigation was warranted, and we set about setting up the date and time of our first meeting. I was excited to say the least and immediately contacted my Tennessee STAR Team to plan our Field Investigation as soon as possible.

Many do not realize the time and money these types of investigations require. My team and I do not charge for our services, and we purchase all of our own equipment necessary for "boots on the ground" investigations. We all have to plan time off of work, coordinate equipment, charge equipment, gather all necessary forms for reporting and legal issues, plan for food, gather clothing (like snake protection, etc., in rural area investigations), and ensure that we are properly armed (this work is dangerous at times as we are in very isolated areas and have no idea what we might encounter). My lead co-investigator, Mr. Don Williams, kindly provides all of our heavy equipment such as heavy-duty trucks for transporting our all-terrain vehicles, numerous equipment bags, drones, and all other miscellaneous tools needed for lengthy investigations. Sometimes we are out all night long in all kinds of conditions. True "hunting" is not for the faint of heart, and many times we endure challenging physical and emotional situations during field work. I'm proud of my colleagues and the bravery and professionalism they bring to our Tennessee MUFON STAR Team!

Once my team was notified about the details of the case, their excitement matched mine, so we set the date of the field investigation as soon as possible. The events that were set in motion from that investigation are still unfolding 3 years later and have come to comprise one of the most interesting cases of my career. We have worked together so long now, I consider Witness "J" and his Mom my family, and I am honored that they have allowed me to investigate, document, and share the events that transpired that night and continue to this day. So, without further ado, let's go deeper into the rabbit hole!

May 21, 2016: First field investigation

When we arrived at the location of the investigation, we were all struck by the beauty of the area. The family farm was originally comprised of over 200 acres of beautiful Tennessee land with a combination of green pastures and heavily wooded areas. The house and outbuildings were immaculately kept, and as we rolled into the long driveway, we were graciously met by Witness "J" and his mother. After introductions, we immediately set about filling out all necessary legal and reporting forms required by MUFON for onsite investigations. I sat down with Witness "J", and one of my other team members questioned his mother separately. Immediately we were struck by the honesty and credibility of each witness. Witness "J" had a military background: E-4P Team Leader Mobile System Switching Operator, Communications Specialist, Secret Security clearance, and 4 years' service with an honorable discharge. His mother presented with: Past Security clearance with the Department of Defense, Naval Air Warfare Center, and Training Systems Division! I just loved her immediately! She is in her 70s, carries a .38 at all times, cannot be coerced in any way, and had no belief in

UFOs or other strange phenomena before all of this transpired. As cases go, things were looking good from all angles of an investigation perspective, and we were off!

Before we go into the details of our investigation, I would like to recap from an investigator's view the events that transpired that night. It really does read like an episode from the X-Files to say the least!

Witness "J" conveyed to me that his father (who also had a military background) had just recently passed away and that he had come to live with his mother from out of state to help with the running of the farm and to inventory certain farm items and tools for possible future sale. As was his current project, "J" was out in the metal equipment building inventorying his father's tools and machinery on a cool spring night. He had the radio playing while going about his duties, organizing and cleaning up the very large work building. Recently he had found a nest of baby birds in the building and had scared off the mother and could not bear to let the babies die, so he had been feeding them with some needle-nosed pliers. The babies were doing well, so he decided to take a video of the nest to share with his mom over coffee in the morning. After taking the video, he thought he ended the video (that's important) on his phone and walked back over toward the side door where his heavy wooden work bench was located. At that moment, he was struck by an odd compulsion to go to the side door. As he approached the door, he laid his phone face down and also took off his sidearm and placed it on the worktable. Both of these actions we will see were strange for the witness as his sidearm and phone usually never left his side. As he opened the door, he was immediately struck by a bluish-white light, and all went blank until he woke up laying on his back on a hard surface. At first, he was really confused upon awakening and could make out a surface above him and the sky. After some time, he was able to realize he was laying on the concrete driveway just under the eave of the house beside the garage doors staring upwards. He had his keys separated between both hands (another strange thing), and his jacket was pulled down around his waist and was somewhat binding of his arms. He was very confused and was having a very strange discussion with himself as he stood up and found his way into the house.

Upon entering the back door through the garage, he immediately went toward his mother's room and found her sitting up in bed almost in a trancelike state. His pit bull was also on the bed with his mother and this detail will be important later as we unravel this case. He called out to his mom, and after completely rousing her, he asked what time it was. She asked him where his phone was and then told him it was about 3:20 a.m. At this point, he freaks out and actually throws his keys across the room. The last memory he had was around 11:45 p.m. or so, and all memory from that point on until 3:20 a.m. was now gone. At this point, his mom was becoming increasingly concerned over her son's bizarre behavior. She asked him again where his phone and sidearm were, and he answered again that he did not know! It was at this point they both assumed the items were still in the equipment building, so they both proceeded to go back outside. His mom tells him that she will accompany him out to the building and cover him with her .38 in case anyone may be around. At that point, Mom is suspecting foul play and she does not want to leave him alone in the condition he is in. They both proceeded outside (it's approximately 3:30 a.m.), and "J" retrieves his phone and sidearm without any other disturbances. On the way back inside, his mom noticed that there were many birds present in the trees, and the birds were excited and making a lot of noise. Also, there was a lot of early leaf fall and other debris scattered about the driveway.

Once back inside, Mom is increasingly concerned about "J" and his odd behavior and her suspicion of some kind of foul play only increases. She asks him to remove his shirt and begins to check him for physical bruises or marks. Once she begins to look, she is now becoming really concerned as she immediately finds what appears to be some kind of strap mark around his chest. Upon further examination, she discovers what appear to be some kind of burn marks behind his ears, a pattern of marks on the top of his head, and some puncture marks on his side. She then documents these physical wounds by taking a series of pictures. (Please note that these, "burn marks, strap marks and puncture sites" were not there the day before). At this point, Mom wants to call the sheriff's office and make a report as she feels her son may have been subject to some kind of attack. "J", still confused, refuses and just wishes to eat and get something to drink. It's interesting to note that he was craving something with salt and was very, very thirsty. These two patterns have turned up in numerous cases after incidents of missing time. Following are the actual case photos!

Photos by Angelia Sheer and MUFON Team. This is how "J" awakened!

Photo by Angelia Sheer and MUFON Team. Notice how the jacket was pulled down which caused the witness to feel partially restrained.

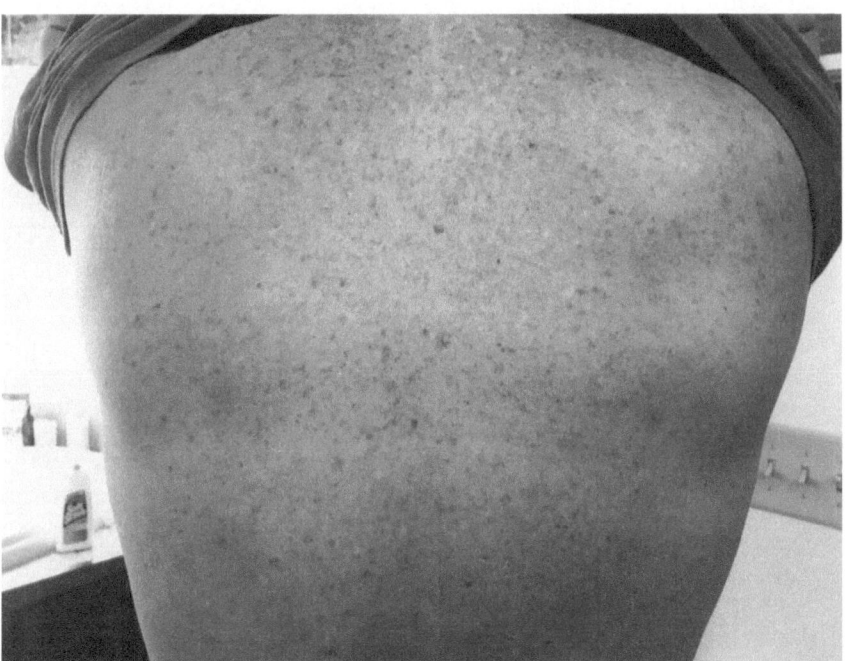

Photos Courtesy of: Witness "J". Author's Notes: Apparent restraint mark across the back.

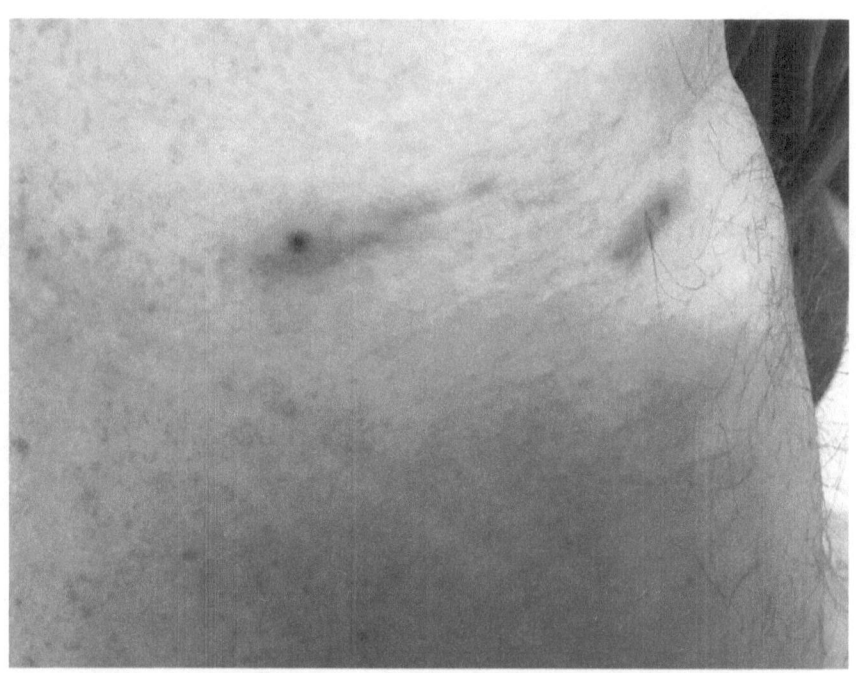

Photos courtesy of Witness "J"

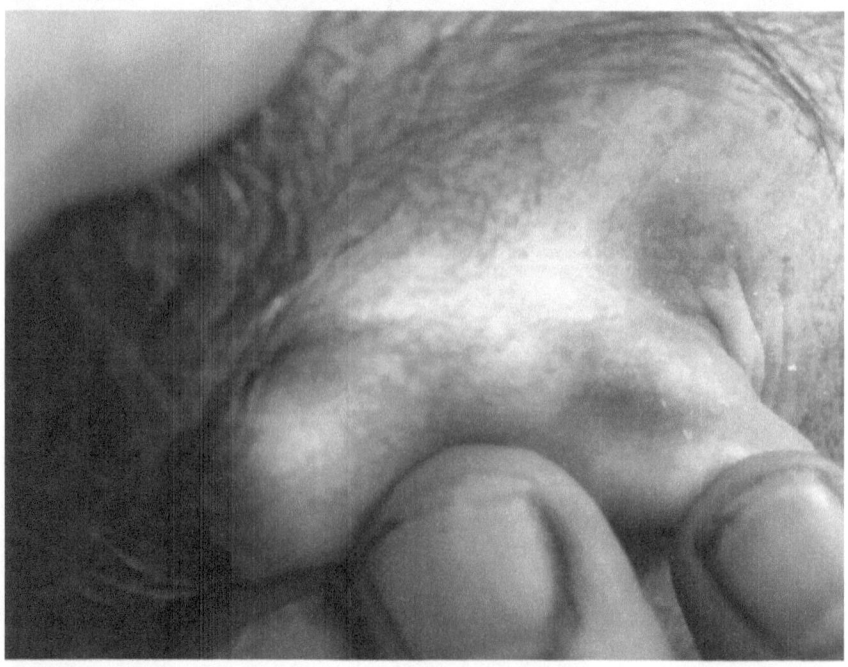

Photos courtesy of Witness "J"

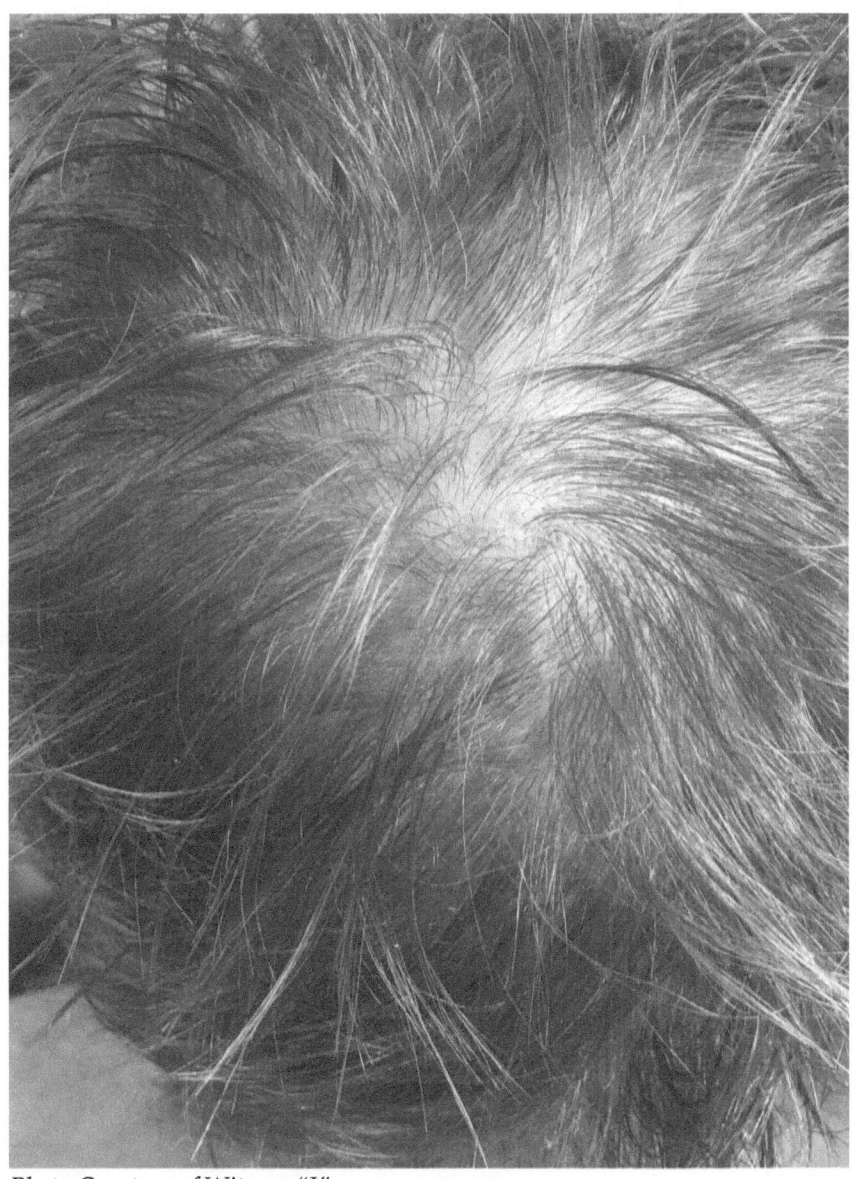

Photo Courtesy of Witness "J"

Once "J" ate something that fateful morning, everyone calmed
down and went on to bed. It was noted that his cell phone was
completely dead, so he plugged it in to recharge and actually did

not think about it until the following evening after a day of work around the farm. That evening when he finally retrieved his phone, he opened the movie that he had recorded with the baby birds and discovered that he had captured 2 hours and 20 minutes of total darkness. Evidently, he had not turned off the phone and had laid it face down on the worktable with it still recording before his missing time event! In that fateful event, he ended up capturing 2 hours and 20 minutes of some of the most disturbing audio I have ever heard.

At this point, my team and I were almost overwhelmed with the amount of evidence being presented. I wanted to listen to that audio but decided on doing a physical sweep of the barn and property first. After finishing up all the necessary paperwork, we headed over to the work barn and started our investigation of the area. We proceeded with EM readings, radiation, and "Bug Detector" sweeps inside and outside of the barn area. Radiation and "Bug Detector" sweeps were negative, but we did get some elevated, stabilized EM readings around the barn area. In my experience, varying EM readings are pretty normal almost anywhere you go, but when these readings spike up and hold constant for periods of time, I pay attention to that! It has been my experience from many case investigations that areas that present with high UFO or high strange phenomena will many times have these high constant EM readings. We also took photos, measured distances from the barn to the house, played out recreations of the night's events, experimented with recordings on iPhones from the work bench area, and moved through our investigation protocols as thoroughly as we could. Next, we unloaded the 4-wheeler. "J" had one of his own, and we all took off to have a tour of the property and wooded areas. At this point, "J" seemed to really lighten up and actually laughed with us as we all had a good time "4-wheelin'" together. This really made me happy as I could tell how heavily all of this had been weighing on him. I think it also was a relief for both of them to share what had happened to date with people who really cared and were serious in trying to find some answers.

The property was beautiful and was comprised of a combination of lush pastures and scenic woods. The lower trails took us off near a beautiful stream down into a valley deeply shaded from the afternoon sun. In one area, "J" had us stop (I think he was wondering if we would feel anything here), and an eerie feeling descended on all of us, almost like we were being watched! He seemed pleased that we had also sensed something just a little off about this area of the woods. I'm glad we actually took the time to explore the farm and woods, because later an event would transpire near this "spooky" location that would add to the overall complexity of this case. My team and I really had a great time exploring and investigating with "J" that day, and I think many other true "boots on the ground" investigators can relate to the absolute joy that comes with the "hunt"! After resting for a bit in a beautiful spot, we all reluctantly climbed back on the 4-wheelers and headed back to the house. We still had a lot to do, and that recording was waiting on us. Little did I know how that one piece of evidence would dramatically affect this investigation as well as all of the people who would listen to it!

Once we got back to the house, "J's" mom had prepared some food for us (God love her!!), so we got freshened up and took a break to eat and socialize a bit. The mood was light, and the conversation was easy. I was glad that "J" and his mom could just relax a bit and talk about all of the weirdness that had been happening to them in a receptive group setting. Even though I could have just sat there and talked most of the night, it was getting late, and we really needed to listen to the audio.

"J" had a private portion of the house, so we all proceeded upstairs to his room to listen to the recording. We all gathered around the computer screen, and "J" explained that he had the original audio but he had also duplicated it and put it into an audio program to help weed out background noises. He warned us that it needed to be loud, so we all hunkered down to listen. The first 20 minutes or so is mostly the radio playing, and then abruptly the music stops (that's another mystery as "J" doesn't remember turning the radio off), and the weird sounds begin. Remember, there is a total of 2 hours and 20 minutes of audio, and during the next 2 hours, we all sat around mesmerized, listening to what I consider some incredible evidence. When we came into the room, none of us had any idea as what to expect. Many people have made incredible claims about their events, but the actual evidence presented usually falls far short of expectations. On first listening, we were all blown away. You can hear "J" moaning, yelling, fighting, cussing, getting sick, and then you hear feminine voices, other voices, and finally a deep, guttural voice that doesn't even sound human. One of the most striking things that happened first was "J's" reaction as the audio really kicked in. He flushed up his neck, his pupils dilated, and his hands began to shake. He had an immediate physiologic response in just listening to the audio, and it was very reminiscent of a post-traumatic stress response. I knew immediately, that _whatever_ happened to this man, it was real, it was terrifying, and it had left an emotional response that could not be faked!

We all just sat there for hours listening, playing back, listening again, and trying to make sense of all that was happening.

Finally, around 10:30 p.m. or so, I knew "J" was exhausted, physically and emotionally, we were tired and overwhelmed with so much evidence to go through and had a pretty long drive home, so we called it a night and loaded up. That was a pretty strange drive home. As investigators, you long for these "up close and personal" cases, and it seems when they actually arrive, you simultaneously want to believe, but in the same moment, something screams out inside that this is just too incredible, too far out, and it can't be true. I am reminded here of my dear friend Debbie Jordon, AKA Kathie Davis of Budd Hopkins' book, "Intruders". Deb poignantly told one of her examining doctors that it was okay if she was "crazy" because "they have a pill for that". The real Experiencers don't want this to be true. As a matter of fact, they would rather be crazy or sick or anything as long as there is some kind of rational explanation. Little did I know that "J" felt the same way, and as time went by and the strangeness just got stranger, he would lament to me that maybe he just had a brain tumor. Unfortunately, I had to remind him that tumors can't make recordings, strange orbs, hovering UFOs, or helicopter flyovers!

Here is a copy of some of the original time stamps from the audio. All of us, "J", his mom, and I have been listening to this audio for almost 3 years. You don't know the nights I would just put on headphones and listen over and over, and I know "J" has listened more than me. This audio has been played on MUFON Radio where numerous individuals reported strange reactions. There is a strange metronome-like ticking that at first we thought may be an artifact of the phone. We did every test we could think of and could not reproduce that ticking sound. There is also a very low frequency embedded in sections of the audio that just seems to come on and then turn off. "J's" distress is apparent throughout, and if that isn't disturbing enough, strange voices waft in, and then eventually what sounds like inhuman voices are heard. The audio ends with the sound of the family dog barking her head off. Remember the dog was sleeping with Mom, and Mom never woke up until "J" roused her from a strange, trance-like state. The dog was barking so wildly, it is heard and recorded some 80 feet away in the barn. The complexity and mystery of this case continues to this day and reminds me that just when you think that you have a small grip on what is happening, something comes along that exponentially increases the questions.

Following is an early time stamp of the audio. Just imagine how many hours went into creating this list!

I use a high-powered stereo system. MUFON uses high-quality headsets and their PC. I suggest using a stereo with the volume almost intolerably loud as I do.

The cell phone was sitting on a work bench. I'm guessing maybe 20 to 30 feet away from me. When the music stops, the doors are closing at the same time, which is impossible to be in two places at once. The stereo is 15 feet from the doors, and you'd have to walk around an 8x4-foot worktable. The voices are within the first hour and six minutes or so.

At 4 minutes, I start talking agitated.

At 9 minutes, I'm talking to someone by side door of barn.

At 13 minutes, I'm answering someone's questions.

At 17:45 minutes, stereo shuts off and rolling barn doors are being shut simultaneously. Strange ticking noise sounds like a metronome?

At 20 minutes, helicopter circling distant.

At 22 minutes, sounds like duct tape? Close to phone and distant banging simultaneous.

At 23:59-24:02 minutes, sound like three-syllable speech unrecognized. I believe I'm drugged at this point.

At 24:20 minutes, sounds like distant people talking.

At 25:35 minutes, I say "\$#* this" followed by sharp command, "Don't touch me!"*

At 25:50 minutes, I say "\$#* you" and then repeat it louder.*

At 26:33 minutes, "I say I'm not..."

At 27:00 minutes, chopper blades.

At 27:13 minutes is when I'm drugged with my distant, indecipherable attempt to speak almost like sounds with a fat tongue.

At 27:19, there's a man's voice over mine saying, "Give him some more." A pause and then   repeats it again

At 27:39, I start fighting back. Multiple low background voices.

At 29:0,0 I slur, "Wha...your...gun...ta...du..?" Pause , distant reply is, "Big red one." Sounds like a distant CB transmission.

At 29:11, female voice says, "Make him stop!"

At 29:20, my drugged voice says, "Ugh...God...please..." I must be in severe pain.

At 29:27, unknown male voice says, "Slippery stuff," then female again makes him stop.

At 29:34, deep non-terrestrial voice speaks god only knows what.

At 29:41, it speaks again then blends to a chit sound (which a sound engineer and I think is possible speech). Meaning we are hearing more unknown speech than we understand as speech. The chirping.

At 29:29, I still won't go out with the drug. I'm trying to fight it.

At 29:46, me, "Don't..." then sounds like getting sick.

At 30:07, me, "Oh..." followed by pain, "Ahh! Haa ha aah!" Then, the frequent noises speed up.

At 30:40, me definitely getting sick.

At 30:46, me in severe pain, "Ah ha ha..."

At 30:53, female speaking in background.

At 32:11, two men conversing.

At 33:40, more air traffic circling. I believe I'm passed out.

At 38:20, I must have gotten sick

At 38:55, someone picks up my ratchet and spins it.

At 43:43, man speaks.

At 44:00 minutes, chopper comes in loud.

At 45:43, man low... Woman murmurs fast.

At.46:00, extremely fast-moving aircraft.

At 47:00, helicopters start to circle.

At 49:00, more predominant, verified by a pilot. People talk in background. At this point, I believe I must be out cold.

At 54:00 minutes, more chopper sounds.

At 1:03:29, heavy bass frequency sound.

Below I would like to include my Case Activity Log to share with readers the time involved in investigating a case and also the struggles witnesses endure in deciding to report a sighting.

Case Log for Witness "J"

5/14/2016:   9:12 a.m. Left message on cell number provided to me to call and discuss case.

*Remember this event happened April 19, 2016, so it took the witness almost a month to report the case. This is a perfect example of the anxiety these people suffer in reporting their experiences. Besides fear of ridicule and the general anxiety unknown experiences engender, many times it almost appears as if some outside agency has "forbidden" the witness to share what happened to them.

5/15/2016: 6:00 p.m. Texted the witness as had not received reply to phone message.

5/16/2016: 6:00 p.m. Phone call, reached witness, initial phone interview.

5/21/2016:   Initial Field Investigation with Tennessee STAR Team. See attached forms and pictures for detailed reports. Basic investigation: Did investigation around the outbuilding, house, and wooded property. Had various low EM readings on field meters (see pictures). Had a very interesting reading on "Bug Detector" frequency range 0-6.0 CHz (see picture of actual meter used). Took extensive statements from witness and mother.

- Entities reported

- High strange activity reported

- Had audio boosted on video file, spent hours listening to content. Various interesting noises can be heard, multiple voices (female can be heard), various moans, distorted background noises, non-human voices heard, overall very disturbing audio.

6/4/2016:

Follow up Field Investigation as ongoing sightings and other high strange events are continuing. Several additional photos taken onsite. Witness and mother report continuing activity that is increasing. Witness and mother continue to send me updates via text. Loud bangs, orbs, increased helicopter activity reported. Witness is convinced that some sort of covert military activity is occurring along with increasing paranormal activity.

6/11/2016:

Ongoing phone conversations, did not want to abandon witness. Very high stress, apparent post-traumatic symptoms, referred to ERT Team for evaluation. Was evaluated and cleared for hypnosis.

6/23/2016:

Witness has developed paranoia to some degree (I really do not blame him as so many things continue to plague him and his family). He asked for me to return his flash drive and all materials. I have complied.

6/23/2016 - 11/9/2016:

Continued phone conversations with witness. Thankfully the paranoia has eased, and we are allowed to have materials returned and evaluated. We are searching but still having no success in finding an appropriate hypnotherapist.

11/2016 - 03/2017:

Phone conversations continue on a regular basis. Finally found a certified hypnotherapist who we feel is a great fit for Witness "J". Hypnosis session performed.

<u>2017 - present:</u>

Witness studies constantly now about what happened to him and is searching for answers. Witness grows stronger each day and has volunteered to appear on MUFON Radio to share his story. He has now appeared on numerous radio shows and has actually spoken at several MUFON meetings. UFO and high strange activity continue to date.

<u>Author's Note: Things to Remember</u>

1) Witness reports paranormal activity from a young age.
2) There are numerous precursive indicators of state changes. The witness had a major episode of missing time. His mother appeared to be in some kind of trance-like state when he went in to wake her up. I find this to be indicative of an outside agency creating the state change when several individuals are affected at the same time.
3) There is some evidence of military involvement as indicated by numerous helicopter flyovers of the farm. Also, evidence could be indicative of advanced consciousness-changing technology either from a terrestrial source or from some unknown outside source.
4) Witness reports that there may be some correlation with drastic emotional changes that he experiences periodically that may herald the advent of paranormal events.

5. FIELD INVESTIGATIONS

"The world is full of obvious things which nobody by any chance ever observes."   *- Arthur Conan Doyle*

Case I: Transforming Lights

MUFON Case Files
Investigators: MUFON of Tennessee STAR (Strike Team Area Research) Team: Angelia Sheer, State Director, Chief Field Investigator; Josh Cross, Assistant State Director, FI; Don Williams, FI, Heavy Equipment
Location: West Tennessee (Exact location is omitted for Witness Protection)
Date: 2017
Witness: "S"

There is nothing better in the field of Ufology than a good old-fashioned field investigation. Cases that actually have enough evidence to mandate an actual onsite visit are rare, so when one arises, my team is always anxious to set out. It's quite a task to prepare for one of these adventures. The day before, there is equipment to choose, charge, and pack, and we make tentative plans for where we will start and end the investigation. We pull maps of the area, taking into consideration the initial location presented, and then create a plan of action to gather as much evidence as possible.

Each of my team members has a very specific role and style of investigating, so as soon as we hit the target area, we disperse and begin our assigned roles. Don Williams, my longtime friend and co-investigator, has been part of my core team from the beginning. He has spent long hours donating his time and heavy equipment for our investigations. By the time he shows up at the rendezvous point, he has loaded his heavy duty truck with ATVs, sighting scopes, coolers of water, chairs, and all the basics we may need for what is usually a very long day (continued into the night) of investigating. Many times, we use the truck as our base camp while using the ATV for transporting people and equipment. Since true Ufology is a grassroots effort, my team and I purchase and maintain all our own equipment, donate our time, and actually assume personal risk in the pursuit of evidence. All of us true "boots on the ground investigators" laugh at the proposition that we are making "all of this money". We all have real jobs, real families, and real commitments that we all must attend to besides our research. I do accept compensation for speaking events and travel but they are truly small and never really cover my actual costs. On average I spend a minimum of 40-60 hours creating power points for lectures, donate my time for regular radio and podcast interviews and never charge a witness for my services. To date a full year has gone into the research and writing of this book and the cover art was commissioned.

I'm very proud of my STAR Team for the time, discipline, dedication, and just sheer bravery they have exhibited in some very strange encounters. The two cases I want to share in this chapter have all the components of extraordinary field work. In these cases, witnesses had reported strange lights, missing time, classic UFOs, entities, and a whole list of paranormal anomalies. I want my readers to get a true feel for what it's like to do real field investigations, so I have included photos that represent a day in the life of true UFO hunters: the work, exhilaration, fear, wonder, and sheer exhaustion of pursuing our passion. I have attempted to choose pictures which will convey to the reader all aspects of what happens on an investigation...so take a look! Then we will get on with two of my favorite field investigations!

MUFON of Tennessee STAR Team from left: Angelia Sheer, Josh Cross, Don Williams. We were on our way to west Tennessee where an "entity' was captured in a video. Photo by Angelia Sheer.

Top: Angelia Sheer and Don Williams "load out". Bottom: Don Williams unloading equipment...here the ramp for the ATV actually broke onsite, and we had to do some really heavy lifting to get it on the ground safely. Photo by: Angelia Sheer.

Case I: Transforming Lights

As state director of MUFON of Tennessee, I have quite a few duties in overseeing the overall administration of the state, but there is nothing more rewarding than opening your email for the day, and there sits a case report that will make your hair stand up. The case was from a young man in west Tennessee who was obviously upset about the activity that had been happening in the area surrounding his home. His case was actually featured as a MUFON Case of Interest. His original report is paraphrased below:

<u>Witness Report</u>

"We have been seeing lights in the sky now for months. Now orange balls of light follow us when we are in our vehicles or on foot. It doesn't happen every night, but sometimes I start to get this weird feeling, and sure enough, something happens. The only way I can describe that feeling is like something is standing right behind you, and it strikes pure fear in you. Yesterday morning at approximately 5:15 a.m., a family member and I went out to investigate the lights in the woods that we have been seeing. When we got to the wood line, we both spotted the lights, and they were moving towards us. Once the light got within 15-20 feet from us, we could see a being standing there. It was gray in color with smooth skin, and it started walking right toward me. The lights that we saw were actually its eyes, and its whole body seemed to radiate a dim light. We were terrified, so we backed off."

Photo by Angelia Sheer   *"Orbs, similar to the ones witnessed in this case"*

The witness reported many nights of seeing these strange lights outside his window in the trees. The lights would move around, change shape, and then all of a sudden, just disappear. The phenomenon was happening on a pretty regular basis, and it had started to really affect him emotionally as well as other family members. Many nights he would go outside, waiting to see if he could capture a picture or video of the lights, and then he would shortly come back inside due to the incredible anxiety and fear that would set in.

The video that was attached to the report was from a game camera that had been strategically placed to try and capture a picture of the strange orbs of light. What he captured was incredible! First there appears a strange rod of light that enters the frame, seems to float about a bit, and then finally touches down on the ground. The moment it hits the ground, it obviously turns into some type of being which walks across the screen between two trees, seems to circle back a time or two, and then disappears behind a tree on the right side of the screen. In a moment, a bright ball of light zooms toward the camera from behind the tree then moves off the screen. It was incredible! I think I have watched that video at least a hundred times. I tried to remove a still from the video but the quality was bad and the entity was not very clear. After watching the video numerous times, I was convinced that it was a legitimate report, and I set about immediately trying to reach the witness. After the initial phone interviews, I was convinced that this was a case worth an actual field event, so I excitedly called my team, and we planned our trip.

It was about a 3-hour trip from home, so after load out, we all excitedly discussed the possibilities of what we may discover during the investigation. As soon as we arrived, we were warmly welcomed by the witness and their family, and we started the investigation with the signing of paperwork and other needed formalities. After all the forms were signed and the initial questioning was recorded, we were all excited to move to the investigation sites. In this case, there would be two primary locations that we would be visiting. The first was the land around the witness's home, and then we would move to the second location where other anomalous lights had been reported.

We arrived at the primary location around 1:00 p.m., and since most of the activity around the home was reported after dark, we decided to go to the second site while we still had light and do our primary scan of the area. The second site was a very rural area that mostly consisted of farms and wooded areas. Along the country road was a bridge where young people congregated on weekends to swim and just hang out. It was at this particular area that my witness and several friends had seen strange lights down along the river. As we started unloading all of our equipment, the first thing that happened was our truck ramp broke as we were unloading the ATV, and it took a good deal of effort to get it down on the ground safely. Once that issue was resolved, we had the witness direct us to the areas in question so we could begin our investigation.

The first thing we planned to do was to fly our drone out over the meandering river where the lights were spotted, so as Josh Cross was prepping the drone, Don Williams and I took off on the ATV to just take a look around. We found a foot path that wound its way down the river that was wide enough to accommodate the ATV, so off we went. We traveled a good piece down the river and didn't find anything unusual, but the farther we went, the atmosphere changed, and a bit of a repressive feeling was noted. We didn't want to hold Josh up with the drone, so after a short while, we turned back and met up with the group in an area near where the anomalous lights had been spotted.

We did a small recorded interview with the witness about the anomalous lights, and then Josh got the drone up in the air. Immediately things started to go wrong. It was a beautiful day around 70 degrees with mild humidity and no wind. Josh had the drone out over the river (Obion River) in plain sight, but all of a sudden, we started to get collision warnings. There was absolutely nothing around the drone that anyone could see. We even pulled out the field glasses and checked, but still nothing could be seen in the path of the drone. Then we started getting warnings of some kind of electromagnetic interference. Up to this point, none of us had ever even seen that type of error or warning with the drone. There for a bit, it was tense as the drone would not respond, and we really thought we were going to lose a very costly piece of equipment. Finally, after a bit of effort, Josh was able to navigate the drone back to the bridge, and we were all relieved for its retrieval. After that, we pulled out compasses and field meters and did detect magnetic anomalies all around the general area. Josh found a spot where the compass would spin erratically. Interestingly, while we were investigating, we were told by several individuals who live in the area about a strange plane crash that occurred during WWII right in this general area.

From "Tennessee History: Tennessee Good Old Days"

https://tennesseehistoryblog.wordpress.com/2017/08/03/palmersville-and-latham-tn-wwii-b-17-crash-sept-1943/comment-page-1/

"Sunday, September 5th, 1943, during WWII, an Army B-17 Bomber crashed between Palmersville and Latham, Tennessee, resulting in the loss of seven airmen's lives. Almost seventy-three years ago to date, the crew, consisting of ten Army airmen, who were flying out of the Dyersburg Army Air Base close to Halls, Tennessee, en route to Gulfport, Mississippi, became lost just after takeoff, fifty miles off course, in the opposite direction that it was first charted. While flying over the northern part of Weakley County, local witnesses stated the plane suddenly exploded midair over the Palmersville and Latham Obion River bottoms. Mr. Hugh Brann of Palmersville, who was only twelve years old at the time, said he witnessed the plane falling while riding his bicycle with friends west of Palmersville. He said "the plane just seemed to come apart as it flew over them" and said he could hear it as it fell from the sky, in what he describes to have been approximately five miles northwest of Palmersville.

According to the Dresden Enterprise, others in the Latham and Palmersville area had also witnessed the plane catch fire and explode and that it had been scattered over a large area between the two towns. And stated some of the wreckage came to rest on, at the time, the Wilkinson, Stowe, and Bondurant farms. According to June Kay Kemp of Cottage Grove, Tennessee, three men had parachuted from the plane, Co-Pilot Second Lieut. Leonard J. Morence of South Bend, Indiana; Bombardier, Second Lieut. Andrew G. Kohihof of Floral Park, New York; and Sergeant Clyde Mullins of Praise, Kentucky, and had survived the horrific crash.

**The newspaper also stated that one of the surviving airmen told a resident in the area of the plane being off course. And at the time of the crash, he thought they were flying over part of Mississippi but was told he was in fact in Weakley County, Tennessee.* The newspaper reported in their September 24th, 1943 issue of the massive amount of equipment that had been brought into Weakley County by the Army to salvage the wreckage and how the Army had posted guards around the area of the crash for weeks while the cleanup was completed.

Mr. Harold Reynolds of Palmersville states he remembers seeing the large equipment coming through town that year and tells about the B-17 crash of 1943. To this day, people, while hunting or logging in the bottoms, tell of finding parts of the plane scattered throughout the woods of the Palmersville and Latham North Fork of the Obion River Bottoms."

Back row L to R: S/SGT Milton Gersfeld Engineer, SGT Clyde Mullins
Radio Operator, SGT Clement J. Funai 2nd Armorer, SGT Donald A.
Goodner Assist. Engineer, SGT Forland F. Nincehelser Gunner and (not

From "Tennessee History: Tennessee Good Old Days"

The actual primary location of the anomalous lights or site #1 was
near Dresden, Tennessee; the second site was closer to the Latham
area along the Obion River. This is where we had electromagnetic
interference with the drone, high EM readings, compass
interference, and, at one point, battery drain on all of our
equipment. It's interesting to note that one of the surviving
passengers of the military crash reported that they were
drastically off course which is highly strange just over farmland
and hints at some kind of magnetic interference with their
navigation equipment. This is exactly what we experienced with

the drone and compass. All of our equipment in this area acted strangely, and at one point all of the batteries drained, and we had to stop and recharge before continuing our investigation of site #2.

When I had first interviewed this witness, there was only discussion of the lights seen near his home. It was not until later that he opened up more and began to describe the incidents at site #2 and then to confide about other high strange phenomena that had plagued him and his family for a while. This is very common in my history of working with Experiencers. Many times, they only reveal the most common aspects of their encounters at first, but as the case moves forward and their trust levels increase, deeper aspects of their stories come forward. This was exactly the case with my current witness, and as the day wore on and we all got to know each other better, he opened up about other events that I could tell really disturbed him. He began to describe an event that happened to him one night while driving home from an evening out. He had been out with friends from about 2:00 p.m. that afternoon and started back to his home somewhere around 11:00 p.m. It was a dark night, and since he lives in a very rural area, no street lights were present, and no other cars were on the road. At one point, the witness begins to see a light following his car. He described the light as an orangish/blue color (remember back to our case with the family stalled on the railroad tracks and my witness' drawing of the lights in front of the car!) that had the quality of LED lighting. The ball of light paced the car at first from about 20 yards away and then slowly gained in proximity until it was right next to the driver window. At this point, the witness describes things as just going surreal. He stated he felt like he was in a movie that just kept repeating a scene over and over again. He remembered hearing an owl (there's that owl connection again!), and his sense of time was completely lost.

Finally, he gets home around 6:00 a.m. feeling completely fatigued, mentally drained, and overcome with a very high degree of anxiety. His trip home should have taken no longer than 30 minutes and yet he did not make it home for almost 7 hours! I could see the real stress in my witness' face as he described this incident, and my sense of compassion for his fear, as well as for all of my witnesses that have endured these events, was overwhelming. These individuals have endured real events that leave real scars with the emotional aftermaths lingering on for years, if not for the remainder of their lives.

After wrapping up our investigation of site #1, we were ready to pack up and move back to the primary location where the original game cam video was captured. It was approximately 11:00 p.m., and this was about the time in the evenings when the witness reported the lights would start to appear. We loaded up and headed back with everyone excitedly talking about all that had transpired in the past few hours. We had already recorded strong electromagnetic interference readings, battery drains, sightings of unusual aircraft, light flares, and the close call with the retrieval of the drone. It was already shaping up to be a great field investigation with everything that had happened so far, but we had no way of knowing what was to come next, and it was one of those events that true investigators can wait a lifetime for!

Overview of findings from Site #2

**Definite EM disturbances were so intense they completely confused our high-tech drone and caused a compass to spin randomly. We lost satellite feed, and the drone kept sending us error messages complaining of EM interference as well as repeatedly going into avoidance mode. It appeared to detect something in its immediate vicinity and took measures to prevent collision, except there was no wind and nothing there that we could detect. We nearly did not retrieve the drone. It was only due to the skill of my right hand, FI Josh Cross, that we got it back.

**High altitude, unidentified aircraft nighttime activity.

**On this occasion, we spotted a fast-moving light that had unusual changes in direction and was unreported on any of our real-time aircraft tracking programs. Of course, there is always the possibility of secret military maneuvers, but the behavior of this object in many ways did not support this.

**At one point in the evening, approximately 10:45 p.m., I witnessed an extremely bright flash of light that remained for a good 45-50 seconds. At first, I thought it might be an Iridium Satellite, but it did not track and only flared that one time. Iridiums have a reflective surface, and when they roll over on their flight path toward the sun, they flare up and then go out repeatedly along their orbit until they are lost to sight. This event happened only once, and the flare was like nothing I have ever witnessed.

Site #1

Arriving back at site #1, we immediately went about checking equipment, charging any batteries that were low, and deciding about our investigation plan for the original event site. We were all really excited about what had happened so far, and even though it was getting late, we were all fired up to get back out in the woods and see if we could actually document those strange orb like lights and entity phenomena. The first piece of equipment I love to bring out for these types of encounters is my FLIR camera. That piece of equipment has never failed me, and it's incredible the detail it can capture.

After deciding which pieces of equipment we would bring, we loaded each other up and made our way through the darkness to the site where the original video and pictures were captured. Earlier, while we had the light, we marked off the areas in question and knew exactly where we wanted to set up to match the original camera position. That night, I had brought along an iPad that we could connect the FLIR to so we would have a nice large screen to track things in. We got everything set up and settled in to see what would happen. It was only a short time before Josh signaled me to come over and take a look at what was happening on the FLIR. We were all stunned...on the screen about 35-40 feet away appeared to be some kind of humanoid entity that seemed to be signaling to us. It moved its arms and legs around, seemed aware of our presence, and we all felt that it was trying to get our attention. I remember to this day, Josh turning to me and saying, "What do we do now?" Needless to say, there are no instruction manuals for these kinds of events. I just looked at him and said, "We punt!" At that point Don Williams volunteered to track out toward the entity through the woods with us calling after him in what direction to head and how far to go. We could not see the entity with the naked eye, but it stayed steady in the camera and just kept on gesticulating in our direction. I cannot describe to my readers the feelings we experienced while standing in the woods watching this entity while it seemed to be trying to get our attention. Don walked on out, and we actually got him very close, and while he could never see or feel anything as he approached the location, the entity stayed steady in the camera. We were shocked, amazed, and thrilled! At no time were we ever frightened or feeling threatened in any way, just incredulous as to what we all experienced.

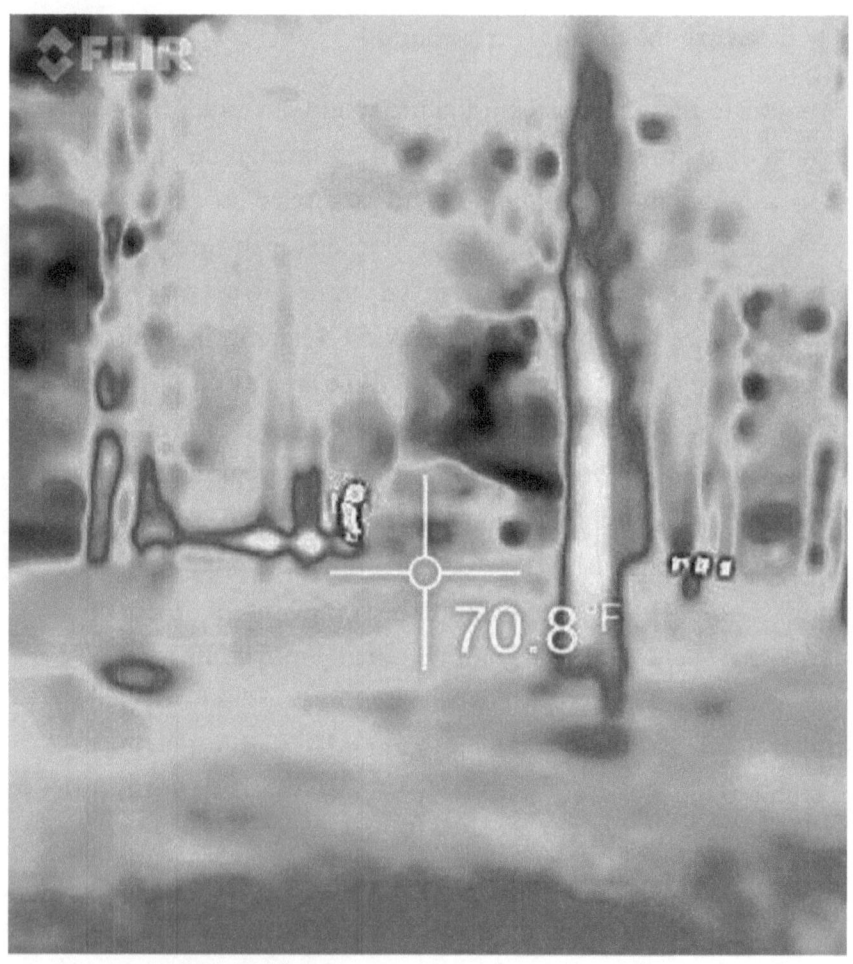

Photo by MUFON STAR Team: Angelia Sheer, Josh Cross, and Don Williams

As you can see in the FLIR capture, the entity is very clear with what appears to be large eyes, an arm, and short legs. Also, notice to the right of the photo the 3 objects just floating above the ground. We have no idea what those were, and we only noticed them later when studying all our pictures. They do sit in the general location where the original pictures were captured by the game camera. Following is a close-up of the entity.

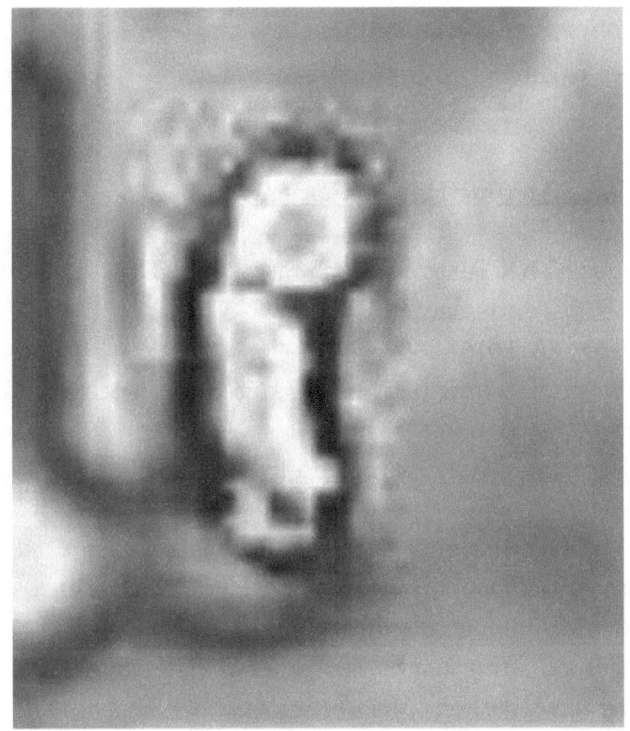

Photo by MUFON STAR Team: Angelia Sheer, Josh Cross, and Don Williams

Overview of Findings from Site #1

**Higher than normal background radiation (not dangerous) and spiking EM readings that were repeated in the same area as multiple UFO, orb, and (what appears to be) entity sightings.

**Actual FLIR capture of what appears to be an autonomous living being that seemed aware of our presence and appeared to be attempting contact of some kind.

**Actual FLIR capture of what appear to be floating, self-illuminating balls of light (orbs).

**The validation of the Witness' apprehensions of changes in the general atmosphere, usually after dark and in the early morning hours prior to dawn, preceding the actual capture of IR activity.

Case II: Does Anyone Have a Camera?
MUFON Case Files
Investigator: STAR Team, Angelia Sheer, and Don Williams
Location: East Tennessee
Date: January 13, 2016
Witness: Samuel (Name has been changed to protect his identity.)

<u>Witness Report to MUFON (Edited to protect Witness identity)</u>

"My wife and I both at night and in daylight have seen multiple bizarre craft. At first, we thought it was from the local underground Air Force base. After watching the object for a time, it appeared to change shape and had bizarre lighting and released other lights from what appeared to be the center lower portion of the craft which later vanished after being ejected. It emitted localized beams of light at homes near us (they didn't go all the way to the ground like normal light but had a place the light actually seemed to stop). My wife ran inside, and I filmed and photographed it and had her load firearms as we were afraid it could be up to something nefarious as its behavior was really strange. The final photo after zoom on iPhone and crop is (seriously) a silver disc with a dome top like a hat. It's not anything like the delta wing three globe light craft we witnessed at dusk like the base tests. I have a notebook of drawings of everything we've seen, and this one is like something from a movie."

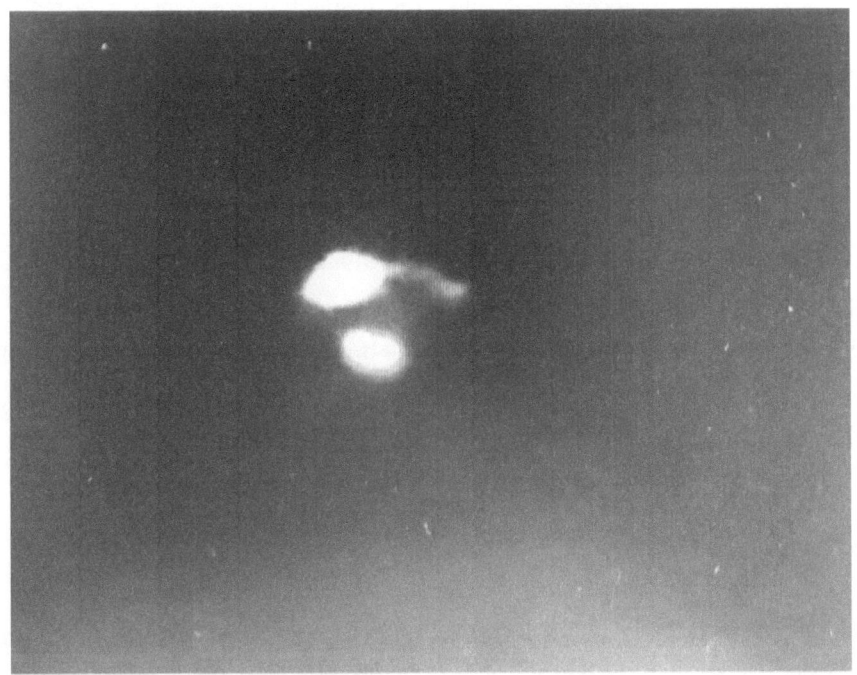

Photo taken from video submitted to MUFON with original report
MUFON media release.

The report contained an interesting video and several other photos that were captured on the witness' iPhone. The video shows a very clear object with a red blinking beacon just leisurely flying over the witness' back yard. In the video, you hear his dog barking like crazy, and as the object continues across the sky, you hear the witness say, "Well, that's not normal". I did a screenshot from the video of the object, and it's really quite a clear picture of a classic UFO. After going over all the videos and photos, I was convinced that this was a credible case and set about contacting the witness immediately. I did reach Samuel in a short time, and he was excited to tell his story, so we set up a time to have an in-depth phone interview.

When I contacted Samuel by phone, we ended up talking for over 2 hours. On most every phone interview, I do record our conversations so I can let the witness tell his story without any interruptions. This way, I can go back later and transcribe the conversation for reports and other investigation needs. Samuel excitedly relayed the happenings of that night and also shared other events that had transpired since. As the conversation unfolded, Samuel shared that he and his family had become frightened with all the recent sightings and had also experienced times when he felt that his home was being watched. At this point, I could feel the mounting stress, so I gently steered the conversation on to the next topic so he could calm down a bit. After completing our initial interview, I was convinced that this case had enough merit to gather the team for a field investigation, so I immediately called the STAR Team together, and we set our investigation date.

I wanted to share this incredible case for two reasons. First, what we experienced that night was one of the most fantastic personal UFO sightings of my life, and it also taught me and my team a valuable lesson about what happens to people in extraordinary circumstances. In the past, I had been doubtful of witnesses who reported close-range UFO events but failed to get a picture, even with equipment close at hand. I would come away from these events somewhat skeptical of the whole story. Well, my team and I were just about to have this lesson come back to us full circle with an event that would haunt us all for some time to come.

The area where Samuel and his family had their sightings was near Lake Normandy, a small but popular recreation area. On our way to the actual investigation site, Don and I decided to scout out the area around the lake as well as take a look at Arnold Air Force Base that is in the same general vicinity. I did some research on the base to rule out experimental aircraft, and it seems there is no longer an active airfield as it was decommissioned in 2009. The base is mainly an engineering development facility at the current time but still sports heavy security as is normal for military bases. We stopped at the guard post and took some pictures with some of the decommissioned aircraft and actually had a chance to talk with some of the guards on duty. When asked if they had ever seen anything strange, of course the answer, unsurprisingly, was no.

After getting a feel for the area, we stopped for a bit to map out our plan of action for the evening and then headed on to the primary research site. When we arrived, Samuel was waiting for us and was excited but nervous about the night ahead. Much of his concern centered on his worry of us not seeing anything while we were there. I assured him that this was out of his control and we were accustomed to long nights of research sometimes with nothing happening at all. He wanted to show us around the property and point out to us where many of the events transpired, so we all set off on a pretty decent hike, discussing the sightings and possible explanations. As we climbed back up the last steep hill back toward the house, the sun was getting low, and it was time now to set up camp before darkness set in.

Don Williams never lets me down on the equipment side of things. We're fortunate to have at our disposal some workhorse pickup trucks and ATVs as well as our general investigation equipment. Once back at the site, we used the truck as the base of operations and went about choosing equipment and cameras and setting up chairs. I like to make these investigations a fun event for the witnesses, so the mood was light as we opened up coolers, broke open the snacks, and just shared good conversation. As the night wore on, I could tell my witness and his family were really having a good time, and everyone was settled in the watch positions, scanning the now black but clear sky. It was around 10:30 p.m. when we first saw it. It came in from the north (it had to have passed over Normandy Lake) and leisurely seemed to head our way. As the object came closer, it was obvious that this was not just a normal aircraft. The first giveaway is that it was completely silent, and as it approached, everything seemed to go silent. The object at this point was right above our heads, and all I can say is we were all dumbfounded. Samuel was having a joyful fit that the craft showed up and was saying over and over, "I told you, I told you." The object was a perfect triangle with 3 white lights on each edge of the craft. In the center, there was a blood-red beacon that we estimated covered about 1/3 the entire size of the fuselage. The object actually came to a stop and just hovered over our heads like it knew we were there. At the point that it hit our zenith, a 747 actually passed over the craft. That was good luck as we were able to calculate an approximate altitude of about 3-4 thousand feet. The craft was so clear to us, we could see what appeared to be windows along the edges that emitted a soft white light. And, being the professionals that we are, we all just stood there with our mouths open in shock and dismay. Not one of us ever thought to take a picture even though thousands of dollars of equipment was scattered about our necks and feet. The object lingered a bit and then gained in altitude, and in a leisurely manner, glided out beyond the trees. But the strangeness was not over, it was only maybe about a half hour later a strange light came up over the southerly tree line. This light was reminiscent of a welder's arch and seemed to dance and skip right above the trees. Its movement was completely erratic, it was silent, and it showed no sign of structure or beacon lights. It seemed to be a free-floating orb of light that fluctuated in intensity, danced along with what could have been interpreted as sentience, and again seemed to tease us with its presence. As I write this, I remember clearly all of our reactions…what do you say to the mysterious, the true unknown? In its presence, you are forever changed!

I will never forget that night…ever! Samuel was beaming that we actually got to see what he had been seeing, and my team had one of the best personal sightings of our lives. The downside… no picture! To this day, I have gained a new appreciation for and understanding of why people fail to get pictures of incredible events. The mind is so overwhelmed at seeing something so out of its domain, it shuts down, and awe and sometimes fear take over. For months after that event, Don would call me and say, "Why didn't we get a picture?" And I'm not exaggerating, for months! It's a joke now, sometimes we just look at each other, and I know what he's thinking about that night. Does anyone have a camera?

Case Update

In July of 2019, Don Williams and I decided to revisit the Normandy Lake area to see if we could once again see our low-flying, triangular UFO. So, on a late Friday afternoon, we loaded up the truck and set off to find a good area for viewing near the dam. This time, undaunted, we had numerous cameras, field meters, "Bug Detectors", FLIR cameras, and any other piece of equipment we thought might be of use. If the craft showed itself again, we were going to make sure to document it. We arrived at the lake just above the dam and settled in on the bank of the lake at one of the public boat launch sites. We had a beautiful spot to watch the sun go down as well as a panoramic view of the lake and sky.

After the sun went down, the area we had chosen ended up having too much traffic, and there were security lights that we hadn't noticed, so we loaded up and went to look for a better spot. Right above the dam, we found a pull off that was secluded, dark, and off the beaten path. The elevation was such that we again had a great view of the lake and the sky, so we pulled all the equipment out and set up our chairs in the bed of the pickup. Don was hysterical with the number of cameras he decided to bring, and we had a fun evening just sitting under the stars, talking about our cases, and keeping a keen eye on the sky.

The first thing that caught our attention was a high altitude, fast-moving craft of some kind that we caught on the low-light camera. We both felt that this was some kind of military craft out for a night maneuver, so we were a bit disappointed that it wasn't our old friend. Later we saw some pretty impressive meteors, but other than that, the night was pretty uneventful. It was getting on toward midnight when we decided to pack it in, but then a really strange thing happened. We were both in the bed of the pickup stowing equipment when a strange, high-pitched tone pierced the area. At first, we thought we might have left some of the equipment turned on, but in all the years I've used that equipment, I've never heard that sound at all. Another strange thing about the tone was that it seemed impossible to determine the source or general direction from which the tone was being emitted. It really was baffling and somewhat disorienting. After a few moments, the tone stopped, and we just shook it off as one of those things and finished up stowing the gear.

After we were back in the truck and had driven just a short way down some gravel roads, the "tone" started again, this time seemingly from inside the truck. I crawled in the back and opened all the equipment bags but still could not find any piece of equipment that might have emitted the sound. At this point, we both were a little unnerved and joked about this being a precursor to an abduction. For a little while, it seemed we might actually be lost on those dirt roads, but after a bit, we hit a main road and wound our way on home.

A few weeks later I was speaking with Deb Kauble, AKA Kathie Davies from Budd Hopkins' book, "Intruders", and I told her about the strange tone both Don and I had experienced during our investigation. I was shocked when she told me about an event that she had experienced during an investigation that was eerily similar. I'm looking into some explanations for that night, but to date, it's still a mystery and a spooky one at that!

6. CROSSOVER EVENTS: HUNTERS, CRYPTIDS, STRANGE LIGHTS

"The mysterious eludes all words of explanation, but it is more significant than all that can be explained. There is always something beyond where we have gone."
                    - Richard S. Gilbert, In The Middle of A Journey

"The Mysteries are gateways, thresholds between this world and the Otherworld, the meeting place of gods and people."
    - Caitlin & John Matthews, Walkers Between the Worlds: The Western Mysteries from Shaman to Magus

Case: I Don't Have a Flashlight!
Parasheer Research Files
Investigator: Angelia Sheer
Location: Union County, South Carolina, Sumter National Forest
Date: January - February 2016
Witness: "Eddie" (Name changed to protect witness identity.)

I love to talk with hunters when investigating strange occurrences in the great outdoors. These individuals are usually no-nonsense people, are proficient with weapons, have spent most of their lives outdoors, are familiar with wildlife in their areas, and tell you like it is. The following witness is just such an individual. I was taken with his honest, calm demeanor and knew immediately he was sharing something that had weighed heavily upon him for a good while. I could tell that he was not prone to embellishment and just narrated his tale to me in a deliberate, forthright manner. To this day, he is astonished at what transpired that dark night, and to this day, he has no rational explanation for the "event". At this time, he lives in a small town and is well known, so he has asked that I keep his identity confidential.

Eddie's Story

On the night in question, Eddie set out with one of his prized 4-year-old hunting dogs and one of his good friends, Sam (assumed name). It was a great night to be out hunting, and a familiar spot had been chosen in the Sumter National Forest. When both men arrived at the location, it was approximately 9:00 p.m., and each man was eager to set the dog loose and let him hunt. The dog in question was named "Max", a well-trained animal who had won multiple field trials and was one of Eddie's favorite hunting dogs. Unlike younger, less-trained dogs that, when loosed, may just go after anything, Max knew his job, and Eddie was confident that he would perform well that night.

As soon as Eddie let Max out of the truck, he noted that the dog's behavior was just off. Max had hunted this area numerous times and had always taken off in the same direction. This night, however, he took off in the opposite direction, tracking into unfamiliar territory. Once again, Eddie paused in his story and stressed to me that things just didn't feel right! Not only was the dog heading off into an area where nothing had been found before, but even the dog's bark sounded odd to him in some way. At this point, things were different but not to the point of alarm, so the men just walked on, waiting for the dog to signal that it had something. It was only a short time thereafter that the dog seemed to have treed something, sounding off with his familiar bark. Both men took off to find Max and wound their way down into a small valley down toward the river. When they finally located the dog, he was up the tree and everything seemed okay, so both men took their positions in trying to determine just what they had. Eddie stayed on this side of the tree with Max while Sam walked around to the far side to take up his position. This particular area was full of pine trees, so as Sam made his way to the other side of the tree, Eddie began to shine a very dim light up into the pines hoping to see if he could get some eye shine. I was educated at this point by my witness that you don't want to shine a bright light up into the tree because the animal has a tendency to turn away from that. But if you shine a dim light you have a good chance of identifying the animal by its eye shine. Well, as Eddie began to shine his light up the tree, he immediately noticed that the upper part of the tree was illuminated with a pretty bright light. At this point he yelled around to Sam and told him to turn off his light. That's when Sam yelled back that he didn't have his light turned on. At this point, Eddie started to look around, reasoning that some other hunter may be approaching the area and wanted to warn the men that he was there so he wouldn't be shot by accident. This was an open area for hunting, and it wasn't unusual to see other hunters from time to time, but this just wasn't the case. Upon closer inspection, both men realized the light source wasn't coming from a nearby hill or pathway but was coming from two balls of light up in the tree canopy. This was extremely unsettling to say the least, and both men paused for a bit, trying to figure out just what was going on. As they watched in stunned silence, what appeared to be two balls of light were moving around in the top of the tree. Since the canopy was pretty thick, it was hard to make out the exact shape of the lights as they moved through the branches of the tree, but both men were sure

159

of what they were seeing. To make matters worse as the two were standing there trying to gather their composure, a terrible smell began to permeate the area, only increasing their agitation. Eddie stressed to me that he had been in the woods his whole life and he had never really smelled that kind of smell before. He struggled to give me a description, comparing the odor to a wet bear, but he stressed there weren't many bears in that area and that description was not exactly right but was the best he could come up with.

This photo was taken at another investigation site but is similar to the type of "orb" light that Eddie saw in the trees that night. Photo by Angelia Sheer.

At this point, things were just getting too strange, so both men decided to get Max and leave the area as fast as possible. As they were walking back, nothing out of the ordinary happened along the trail, but as the men approached the access road where the truck was parked, the balls of light once again reappeared. On the lefthand side of the road, there was a stand of pines, and now the lights were in front of the two men in the top of those pines. Both the lights moved around again in the top of the tree, and then they both jumped over the access road to the top of another tree. The witness could not see an actual beam of light coming from the objects, but the light the objects were emitting was enough to illuminate the ground below the trees. Eddie estimated that the balls of light were about the size of a softball and were a silver-white color. Both balls of light stayed in the tree for a short time and then shot straight back up into the sky.

After the experience, both men talked about what happened, and Eddie stressed to his friend, "You saw that right? Don't ever forget what happened here." His friend acknowledged seeing the strange lights but after a short time refused to discuss it again. I asked if I might interview his friend, but my witness stressed to me that, to this day, his friend is very hesitant to talk about what happened to them that night. Unfortunately, this is very common as individuals try and deal with the unknown and their fears of not only what happened but how they may be received in sharing their event. It was over a year in which Eddie and I got to know each other, and it was only in time that he felt comfortable to come forward and share not only the events of that night but what he began to uncover about himself. It was only with time and retrospect that he began to realize that numerous other strange happenings had followed him his whole life.

Eddie describes the following details that fall completely in my tracking of common denominators shared by witnesses. In no way do I ever lead a witness to these reports. Instead, I let them recount their stories in the time needed to unfold. As more details come forward and each witness begins to retell so many similar events that I have heard from thousands of witnesses over the years, I always smile to myself and reiterate my favorite saying, "You just can't make this stuff up!"

Eddie reported the following:

While standing outside in his yard one day, an owl flew down very close to Eddie and just stood there, watching him in a very unnatural way. From this experience, he began to remember seeing owls in his window as a young boy. As he was sharing this, I could tell that all of this was unsettling, and I could begin to see the start of retrospect memory integration. This is where many, many strange experiences begin to flood the witness' consciousness, and the realization strikes that these things are real and have been happening usually most of their lives.

The witness has a hunting cabin in a very remote location and enjoys spending a lot of time in the outdoors. One night while staying at his cabin, Eddie witnessed and photographed what he feels is some kind of paranormal entity. This event was so disturbing that it prevented him from visiting the cabin as much as he used to. He stressed to me that he had spent his entire life out in the woods and had never been afraid. After these events, there are times at his cabin that he feels he is being watched, and the overall atmosphere is quite disturbing.

Eddie reports to this day that he is extremely sensitive to others' emotions and is many times overwhelmed and will seek out places of solitude to recharge. He shared with me that all his life, people have come up to him and readily talked with him about the most intimate details of their lives. They feel safe confiding their troubles with him, and he always seems to calm them down, and they leave feeling better. This happens to many of my talented, sensitive witnesses, and although this is helpful to the person sharing, unchecked without some knowledge of shielding can lead to exhaustion and overwhelm. (This, by the way, is one of the quintessential signs of a gifted Empath.)

Author's Notes – Things to Remember

1) Pay attention to how one startling event brought back the integration of numerous memories that had seemingly been blocked for years.
2) Notice how other paranormal events have seemingly manifested throughout the witness' life, i.e., owls in the window and sightings of possible strange ephemeral beings.
3) And finally, pay attention to the fruition of all these events. The witness develops some kind of keen sensitivity (in this case, sensitivity to others' feelings, etc.), and a profound Transformation of Consciousness overtakes their life.

"Bigfoot" and the UFO Connection

About 3 years ago I exhausted the UFO literature (I'm usually reading about 3 serious books at a time across the board in the fields of ufology, science, psychology, trauma, etc.) and just wanted something fun to dive into. I quite accidentally discovered the fields of "Cryptozoology", which is the study of strange animals purported to exist. As I do with most things, I was obsessed and read everything I could find, scientific and otherwise. The first book I chose was Jeff Meldrum's "*Sasquatch: Legend Meets Science*", and wow, was I blown away. First of all, Dr. Meldrum's bio reads as follows:

Biographical sketch from Dr. Meldrum's website:

"Dr. Jeff Meldrum is a Full Professor of Anatomy & Anthropology at Idaho State University (since 1993). He teaches human anatomy in the graduate health professions programs. His research encompasses questions of vertebrate evolutionary morphology generally, primate locomotor adaptations more particularly, and especially the emergence of modern human bipedalism. His co-edited volume, From Biped to Strider: The Emergence of Modern Human Walking, Running, and Resource Transport, proposes a more recent innovation of modern striding gait than previously assumed. His interest in the footprints attributed to Sasquatch was piqued when he examined a set of 15-inch tracks in Washington in 1996. Now his lab houses well over 300 footprint casts attributed to this mystery primate. He conducts collaborative laboratory and field research throughout North America and the world (e.g., China, Russia) and has spoken about his findings in numerous popular and professional publications, interviews, television and radio appearances, public and professional presentations. He is author of Sasquatch: Legend Meets Science (Tom Doherty Publishers), which explores his and other scientists' evaluations of the contemporary evidence and also affords deference to tribal people's traditional knowledge of this subject. He has also published two field guides, one focusing on Sasquatch, the second casting the net more broadly to consider the potential of relict hominids around the world (Paradise Cay Publishing). He is editor-in-chief of the scholarly refereed journal, The Relict Hominid Inquiry."

To have someone of Dr. Meldrum's credentials write an entire book on the scientific evidence for the existence of "Sasquatch" was incredible. At this point, I was really hooked, and down the rabbit hole I went. After finishing Dr. Meldrum's book, I devoured everything I could find and was astonished at the sheer volume of reports.

Another honorable mention goes out to David Paulides and his book, *"The Hoopa Project"*. Mr. Paulides, a former police investigator and researcher extraordinaire, is also known for his *"Missing 411"* series. I was again amazed at the sheer volume of credible witness testimony of Bigfoot sightings. Also unbeknownst to me, I discovered actual legislation that has been enacted concerning this elusive creature. From "The Hoopa Project":

"An Internet search for Bigfoot/Sasquatch legislation came up with Whatcom County, Washington. I call the county council and was fortunate to be assisted by a very friendly county employee. I explained my request and asked if she had ever heard of the legislation. She stated that she had but had never received a request for a copy. She said that she would gladly send me a certified copy. Their legislation was enacted on June 9, 1991. The bill calls for Whatcom County to be declared a "Sasquatch protection and refuge area".

Why in the world would any state or local government spend the time or budget to pass frivolous laws based on a myth? That just doesn't make any sense at all. Also, a fellow researcher's daughter was enrolled in a survival course in Washington State that was funded by the military. During her training, she reported that in her textbook was a section on animals to be avoided. Guess what was listed. It warned to avoid Sasquatch at all costs, do not engage, and leave the area immediately. I am currently seeking a copy of this document. Again, if this is all a mythological animal, why include it in a serious survival manual?

Find following a copy of the Ordinance that is mentioned in "The Hoopa Project" referenced above. Very interesting!

bigfoot.res 6/9/91

SPONSORED BY: _Consent_____
PROPOSED BY: _Harris_____
INTRODUCTION DATE: _6/9/92_____

1   RESOLUTION NO. _92-043_

2   DECLARING WHATCOM COUNTY A SASQUATCH PROTECTION

3   AND REFUGE AREA

4   WHEREAS, legend, purported recent findings and spoor suggest that Bigfoot may
5 exist; and

6   WHEREAS, if such a creature exists, it is inadequately protected and in danger of
7 death or injury;

8   NOW, THEREFORE, BE IT RESOLVED by the Whatcom County Council that,
9 Whatcom County is hereby declared a Sasquatch protection and refuge area, and all
10 citizens are asked to recognize said status.

11   BE IT FURTHER RESOLVED, this resolution shall be effective immediately.

12   APPROVED this _9th_ day of _June___, 1991.

13   WHATCOM COUNTY COUNCIL
14 ATTEST: WHATCOM COUNTY, WASHINGTON

15
16 Ramona Reeves, Council Clerk   Daniel M. Warner, Chair

17 APPROVED AS TO FORM:

18
19 Civil Deputy Pros. Atty.

I actually met up with David at a conference last year, and along with Travis Walton (*"Fire in the Sky"*), David Bakara (*Expedition*

Bigfoot), and Dale Houston (fellow researcher and country music singer), we had a most intriguing dinner together, discussing our respective fields. Into the night, we talked about many of our cases and all of the high strange events that weave in and out of credible witness testimony and available evidence.

So as my "Bigfoot" education progressed and I moved from one book to another, page after page, I was shocked at the similarity between cryptid and UFO sightings reported by very credible witnesses. There were law enforcement personnel, park rangers, doctors, lawyers, housewives, and the list went on. As in UFO sightings, most of these people had nothing to gain in reporting their events and everything to lose, and yet, year after year, the reports kept pouring in. As my research took me further and further into the existing literature and actual witness testimonies, I was convinced that some type of upright, possible proto-human creature was stalking not only the northwest United States but, astonishingly, was being reported all over the U.S. and around the world. It was at this point that I came across the work of Stan Gordon, referenced in earlier chapters, and just could no longer ignore the connections that just kept cropping up. To remind readers, Stan Gordon documented over 276 UFO/Bigfoot related sightings between the years of 1973-1974 in the Pennsylvania area. The local police department worked with Stan and his team in relaying reports to him, and several officers were involved in actual sightings. There was one case in particular that Stan reported in his book, "*Silent Invasion: The Pennsylvania UFO-Bigfoot Casebook*", that really convinced Stan and his team that something extraordinary was going on. An object that appeared at first as a red glowing sphere witnessed by over 15 people landed in a pasture on a local farm. Subsequently, two Bigfoot-like entities were seen walking toward the witnesses, apparently from the landed craft. What ensued after that is like an episode from the X-Files, and I urge you to get Stan's book and dig in.

Now, as synchronicity would have it, I was invited to speak at The Tennessee Bigfoot Conference in October of 2018 in Kingsport, Tennessee. I was honored to be asked but told the promoter that I was mainly a UFO Researcher and was concerned that my information may not fit in with the other speakers. He would have none of that, and so off I went to speak at my first Bigfoot convention, and wow, did I have fun! Not only was the audience into the UFO part of this phenomena, but I had many individuals wait around after the lecture to share their own stories of sightings and strange events. Furthermore, it was there that I met my now dear friend and fellow researcher, Mr. Matt Delph. Matt was one of the keynote speakers and was very well known in the Bigfoot Research world. As we got to know each other and I heard his incredible story firsthand, I was convinced of his sincerity and the legitimacy of his encounter. I interviewed Matt, and he shared his incredible story with me:

Matt's Encounter

Matt, now 44, grew up in the countryside of Lafayette, Indiana, near the junction of the Tippecanoe and Wabash Rivers. His parents owned about ten acres that was surrounded by hundreds of acres of forest land where he lived with his siblings and enjoyed the country life. The house was completely isolated at the end of a gravel road in the middle of a dense wooded area. In the back of the home, there was a small creek that eventually found its way into the Wabash River, so water was plentiful and so was a multitude of wildlife. The family raised chickens and hogs, so there was always plenty to keep the children busy, and in their spare time, there was plenty of land to play on. Beyond the creek, the family maintained a cleared area, and where the flatland sloped off, a hill rose nearby that played host to a growth of raspberry bushes and vines.

Over the years, Matt, his siblings, and their parents had noticed areas that had been pressed down as if someone or something had recently slept there. Other odd things would sometimes happen, but just as most of us rationalize away strange things, so did the family. Many times, they would notice strange odors coming from the flattened areas or feel as if they were all being watched from a distance. During those times, the dogs would go crazy barking until they seemed to drive "something" away, and the feeling would fade. In 1992, sometime in the late summer, the family heard a loud scream that could not be readily identified. The "scream" seemed to come from an area very close to the house. Again, the dogs went crazy, and Matt's mom and two sisters became very upset. Eventually, the barking dogs seemed to drive away whatever had caused that terrifying scream.

About a month had passed when another strange event took place near the family's home. Matt's siblings and a friend were playing in the woods behind the house when they started to hear movement in the brush. At first, they thought it was just a deer, but again that strange putrid odor began to fill the air. Being kids, they all decided to throw rocks into the area where the odor seemed to be coming from, hoping to get a glimpse of the animal.

From this point on in the story, Matt reminded me that he was not present for this event, and he referred me to his sister, Amanda Tolbert. I was excited to speak with Amanda about what happened that day as this brought in another witness to these strange events. While we were talking about her experience, Amanda shared that her experience was sometime in the late summer, probably in August, while Matt's sighting happened the week before Thanksgiving about 4 years later. This again sets up a pattern of events that had been happening to the family for many years but only became clear in retrospect. The following is Amanda's story:

"When I was about 9 years old, my other brother and a neighbor friend decided to go back behind the house in the woods where we had a lot of trails to play on. The woods were dense and led down to a creek that we would also play around. I want to stress that behind our house, there was nothing but these dense woods and brush and a creek that eventually led down to a river. When we got down on the trails across the creek, we started to hear a lot of movement in the woods and also started to notice a very strange odor. Thinking that we were just hearing the movement of some deer, we started throwing rocks and sticks into the area where we could hear the movement through the brush. This is when we heard a very loud, terrifying howl come from the woods. Then, whatever was in the woods actually started throwing sticks back at us. It was at this point that I actually got a glimpse of the thing that was howling and throwing the objects back at us. I saw the shoulder and part of its arm as it threw out one of the sticks from in the brush. The thing was huge, around 7 feet tall, and a dark brown in color. Needless to say, as children we were terrified and took off full speed running back down the trails and into the woods. At some point, we did get separated as there were numerous trails for us to use, but eventually we found our way to an open corn field that was past our neighbor's house. Here we were finally able to regroup and then made our way out of the cornfield back onto a well-used back road. At this point, it was getting dark, and we were about a mile away from our home and about half a mile from our neighbor's house. We then proceeded to walk our neighbor back to his

house and then made our way on home. When we got home, we tried to tell our mom what had happened, but she was so upset with us for being gone so long that she really didn't hear what we were trying to tell her.

In thinking about this event, it has become clear to me that strange things had been happening around our farm for many years. There were many times we heard strange howls, movement in the woods, and noticed that strange odor, but we always just rationalized it away as some kind of normal animal, because in those days, no one ever talked about the possibility of a Bigfoot. Furthermore, as I think back about that eventful day, I remembered numerous nights where I would actually see something walk by our windows in the darkness. I would go get my Dad and tell him someone was walking around the house, and he would diligently go outside to check on things, but he never would find anything. This happened numerous times throughout the years, and I am now convinced that it was one of these creatures. Also, as I think back on all of the events that happened to our family over the years, even though we were sometimes really scared, I don't feel that these creatures actually meant to really hurt us. It really just felt like, this is my area and you just need to get out of here! - Amanda Tolbert

Again, I want to stress the high credibility of this witness as well as her family. The direct presentation of the facts of the event without elaboration and the overall links with Matt's encounter make this a very strong case. I not only see this in many of my paranormal cases but also in the accumulated UFO cases that I have researched. There are very high percentages of other family members reporting lifelong strange events that, when considered in retrospect, show clear patterns of some kind of intelligent interaction. Amanda also mentioned in the closing of our conversation that she felt that these beings were "just playing around with us!" That's a very interesting statement as well as telling. Animals may stalk humans, but they don't use that kind of intelligent design in manipulating their prey. So, back to Matt's story:

As life will have it, the years drifted by, but Matt's love of the area in which he grew up would always draw him back, and he would return periodically to visit and go deer hunting. Matt is an avid outdoorsman, tracker, and skilled hunter, is and always on the lookout for that perfect buck. As fate would have it, Matt had returned home for a visit the week before Thanksgiving Day in 1996. He was happy to have some time off and planned to spend as much time as he could out in the woods, deer hunting. He excitedly gathered up his gear and headed out alone, starting down the trail into those familiar woods. Just like in his childhood, he had no problem finding his way down to the creek and cut a path along the bank which was low and somewhat marshy. As he wound his way through the brush, he came to a point where two deer trails actually crossed each other. He decided this would be a good place to settle in and wait to see what came down those trails. He found a good tree to sit next to, and he went about clearing away the leaves in the general area so as to not make any unnecessary noise. Hunting is a huge waiting game, and he knew he might be here for a while, so he made note of the time, 2:30 p.m., and settled in to watch the sun make its journey across the sky. Many hours had passed, and sundown was fast approaching when he was alerted to the sounds of some kind of animal moving through the woods. The rustling was definitely getting closer by the minute, so he immediately prepared his weapon in the hopes that a deer was on its way toward him.

Immediately he was struck with that eerily familiar terrible odor that he and his family had experienced so long ago. None of them had ever really figured out what had created that smell, but as so many of us do when faced with the unknown, he and the family had just rationalized it away. In the excitement of the moment, the best he could come up with is that a buck was tracking a doe in heat and shortly they would both come crashing out of the woods, providing him with a great shot.

Normally deer are quiet in the woods with the exception of two bucks competing for a female. Often in these cases, the bucks make quite a ruckus in contention for the doe. As the sounds drew nearer, some small sapling trees in the distance began to shake in a very disturbing manner. It was only about a hundred feet in front of him, and this only increased his excitement in the moment. What happened next would change Matt's life forever. In an instant, there was the most terrifying, blood-curdling scream combined with a growl that he had ever heard. Matt was immediately stunned as there was no mistaking the territorial and obviously aggressive nature of this scream. Instinctively, he readied his weapon as it became very clear that something did not want him hear it. It seemed that time really slowed down as it does in these moments, the crashing continued on unabated, and the tension levels just kept increasing as whatever it was drew closer and closer.

Then it happened…Matt saw something move through the clearing between a gap in the trees. The creature that he spotted was like nothing he had ever seen in all of his years spent outdoors. It was bipedal, between 6-7 feet tall, had dark hair, and at this point was less than 30 yards away and was increasing its aggressive stance. It was at this point the creature moved from left to right from one grouping of trees to another in an angle still approaching the terrified hunter. While it approached, Matt could very clearly make out the movement of a very broad shoulder and part of its arm. At this point, a very large branch came hurtling out of the woods directly toward him. He actually had to duck to avoid being hit by the large limb! The creature was clearly aggressive and let out another terrifying scream accompanied by a deep, low growl.

For those of you who don't know Matt personally, he is one of the bravest individuals I know. Many nights, I actually worry about his safety as he is out alone all night, researching. He admits in this case he wasted no time in getting out of there. He says he wasn't sure how far he ran before he stopped to look behind him and reassess the situation. As far as he can remember, he was halfway down the hill and breathing heavily before he came to a stop, staying stock still and listening for any pursuit. For a brief moment, all seemed to be clear when came another awful scream combined with those terrifying, threatening growls.

In a moment, he was off again. He cleared the marshy areas along the creek and didn't stop until he reached the safety of his parents' home. Of course, no one was home, so he didn't wait around for any more interactions, he jumped in his truck and headed to his home in West Lafayette. On the drive, he had tried to calm himself, so when he arrived home, he would not raise any suspicion from his roommate. He must have not been too successful, for his roommate immediately knew something was wrong, and after some prodding got Matt to open up about his unbelievable encounter. The roommate believed Matt immediately as he knew Matt's character, and both agreed they wanted no part of those woods behind Matt's parents' house for the time being.

Speaking with Matt now, he laments how things become very clear in retrospect. The sounds, odd noises, odors, and mysterious screams of his childhood take on a whole new meaning. So many things that were ignored or just rationalized away become clear patterns of behaviors reported in Bigfoot sightings all over the country and the world.

Ever since that fateful day in 1996, Matt has made it his mission to research that amazing creature that burst out of the brush and threw a stick at him all those years ago. An experienced outdoorsman, hunter, and tracker, Matt is proficient in recognizing the habits and patterns of behavior of all types of animals. He currently resides in southwestern Virginia and frequents the dense forests, looking for evidence of that elusive creature he encountered so many years ago. Matt is also one of the original founders of MECRO or Mountain Empire Cryptid Research Organization where he shares the findings of his research freely. You can also catch him live at one of the many Bigfoot Conferences around the southeast where he gives talks about his experiences.

In the last few years, Matt has become one of my dear friends and has ushered me into the "behind the scenes" Sasquatch research. Again, the sheer number of reports from hunters to housewives continues to grow as does the interest in this mysterious cryptid. I cannot go into all the documentation, DNA research on samples (check out Melba Ketchum's work), or the huge volume of credible documented sightings, but it is worth your while to do your own research. I am forever thankful for Matt's knowledge, patience, and especially his sense of integrity. I wanted to include his story here even though, to date, he has not experienced a UFO sighting personally, but he has interviewed many individuals who have reported some really strange things. Cryptid witnesses continue to report strange orbs in trees, actual UFO encounters, and other paranormal events in conjunction with their sightings. All of us who have been out in the field long enough, no matter our area of expertise, come to discover the strange crossover events that permeate our research. I and many others are working on collaborative projects, so we may share our findings and compare notes. I feel through these collaborative efforts, our ability to find answers will increase exponentially.

A note from Matt…

"I have come to know and work with Angelia over the past few years, and the amount of crossover evidence in our respective fields is truly amazing. By sharing our witness accounts and field research notes, it is incredible the number of similarities that crop up in mutual reports. Strange lights, floating orbs, and events of high strangeness show up in my cryptid research just as cryptid sightings now permeate Angelia's UFO investigations. There is a connection and a big one. By working together, it's amazing the things we have tied together, and each day, with each report, we move one more step closer to the truth." - Matt Delph

Notes from the Author:

Matt and I have discussed in detail the state change issue in witnesses, and he now watches for signs of these events such as missing time, confusion, memory distortion, and time anomalies with his witnesses.

We have also discussed in detail the number of times he actually wrote off genuinely strange events in his life as just something mundane. But after his own personal sighting, he began to piece together other events of significance in the past that indicated that the phenomena had been with the family a long time but just rationalized away. You don't know the times witnesses will say that they have only experienced that one strange incident, but after some time and reflection, other memories come flooding back that prove that is inaccurate. Many times after the initial memories are questioned during a formal interview or with hypnosis, the flood gates are opened, and an ordered pattern of contiguous phenomena emerges.

Angelia Sheer and Matt Delph doing what we love to do…sharing our work and research at a conference!

Case: Bright Lights and Missing Ammo
Source: MUFON Case Files
Investigator: Angelia Sheer
Location: Gatlinburg, Tennessee
Date: January 14, 1969
Witness: "Mac" (Assumed name; witness desires to stay anonymous.)

Mac's Account

"*A group of four of us went on a raccoon hunt one winter night in 1969. We crossed a field and went near the top of a mountain where the dogs were trailing, so we sat down and turned our lights off while the dogs went off doing their thing. We were sitting around on the ground, and some of us were sitting up against trees, cutting up like teenagers do, when all of a sudden, a large, super bright round light came on like someone flipped a switch. I remember standing up and looking up, and it was directly above us. The light was so bright that I couldn't make out any shape of an object it was coming from. The light had to be 100-150 feet in diameter if not larger. Someone yelled "run", and I ran for several hundred feet and turned around, and the light was gone.*

We all gathered back to where it happened and asked one another, "What the hell was that?" None of us were able to identify the source of the light, but we all agreed there was no sound. We went on with our hunting trip and stayed out the rest of the night. The next day, my cousin who was with us asked me to go back and help him find the tubular magazine for his .22 caliber rifle that he lost that night. I tried to talk him out of going to look for it, saying we had traveled miles so it would be impossible to find it. I finally said okay. When we got to where we saw the light, we found the magazine along with all the .22 caliber bullets on the ground in a pile. One of the boys that was with us that night later became a policeman in Florida, and after he retired and moved back here, we talked about that night, and he did not remember any more than I do. Do you think hypnosis would help me recall things that might have happened after we saw the lights?"

I spoke with the witness by phone on February 21, 2017. The witness was excited to speak with me about his experience and also curious about memories that had just recently returned surrounding the incident. One of his friends that was in the original group had just moved back to the area, had remembered the incident quite clearly, and also remarked that the memory surrounding the events had just recently returned. This was extremely interesting that two separate witnesses had memory recall around the same age and so many years later. Mac expressed to me that he felt like more happened to him and his hunting companions that night but, to date, this was all he could remember. He asked me if I thought hypnosis might help, and I told him that in many instances, it is a great tool. Here is an excerpt from my actual interview notes:

Phone interview 2/21/2017: 1969, December-January, best guess, hunting, about 9-9:30. Ran to his right, went about 20-25 feet behind a tree, and when he turned around, it was gone. No sound, no wind. There were 4 witnesses. Brother and two friends. The shells being on the ground were strange. They kept on hunting late into the night. About 4 years ago, a friend came back from Florida; he asked him about the lights that night, and he remembered it.

Altitude of lights: 150-200 feet, best estimate

Radius that covered: 100 feet diameter in area.

Winter time: Skin was covered. I asked the witness if he felt any sensation on his skin from the light. He does not remember feeling any heat or other sensations.

Physical Effects: Was some nausea that night. No loss of hair.

Psychological Effects: He has wondered over the years if something more happened to him and his companions that night. From his perspective, it seemed that they all wandered in the woods for a very long time, but he cannot be sure. No anxiety developed around this experience. To date, he has not spoken with his brother or friends about the incident.

I suggested to the witness that he contact the others involved in the sighting and compare memories. I also recommended he ask them about the ammunition event. Told Mac to call me anytime if he remembers anything else and check in with me if he speaks to the others. I told him I would call him back in about 2 weeks for an update.

After our last phone interview, I lost touch with the witness. I was hopeful that some of the other men that were present would come forward and relate their memories of that fateful evening. Mac was able to speak with one of the men there that night, and he did confirm that he remembered what happened, but like so many others, he was reluctant to talk about it. I am from the south and was raised in a family of hunters. No hunter that I have ever known would willingly open their guns, take out their ammo, put it in a neat pile, and then just walk away. The witness stressed to me that the boys didn't have that much money when they were young and would never just leave their ammunition anywhere. Matter of fact, that's why they returned the next day to see if they could actually find it. Mac related to me that he was actually surprised that they found the location and were able to retrieve the ammo.

I asked him when he returned if they noticed anything unusual at the site, but he said they didn't and were just happy to find the ammunition still sitting there. We talked for a while longer, and I could not discover any more details, but Mac stressed over and over how this memory just came back to him recently and that it just nagged at him and that's why he decided to make a report to MUFON. We talked about hypnosis, and I told him I would check and see if there was someone reliable in his area, but I never reached him again after that phone call. Like Mac, so many people have contacted me over the years with memory loss and fragmentation, and I had so little luck finding them a qualified hypnotherapist, I finally decided to seek training. I have a nursing background, and in many cases, had already built a rapport with so many of my witnesses, it just made sense. It took some time to find a program that I felt was up on all the most current neuroscience and had graduated practitioners who displayed outstanding results. After almost a year-long search across the country, I found an extraordinary teacher right here in Nashville. I am now happy to be able to provide sessions for my witnesses who feel they have missing time or have a nagging sense that something more has happened to them.

Case Update: I loved this case from the start as I know hunters, and I know they would NEVER leave their ammunition just sitting on the ground in a neat little pile and then go wandering off into the woods hunting again with NO ammo! So, I decided to try and find my witness one more time, and luckily, I did manage to get in touch with him. He was really happy to hear from me, and I shared with him that I would love to include his story in this book. He was thrilled and agreed to help in any way he could. To his knowledge, there is only one surviving witness left, and he volunteered to see if he could locate that individual and see if he would be willing to speak with me. As of this date, the witness tried repeatedly to find the other individual that was present that night but has been unsuccessful. Hopefully there will be a future book, and if there are any updates, I will keep you posted.

7. NOTEWORTHY PARANORMAL EVENTS

"All a skeptic is, is someone who hasn't had an experience yet."
- Jason Hawes

"Surely the supreme problem for science to solve if she can, is whether life, as we know it, can exist without protoplasm, or whether we are but the creatures of an idle day: whether the present life is the entrance to an infinite and unseen world beyond, or "the Universe but a soulless interaction of atoms, and life a paltry misery, closed in the grave."
- William F. Barrette, On the Threshold of the Unseen, 1918

<u>**Case: Reading Minds: The Case for Telepathy**</u>
Parasheer Research Case Files
Investigator: Angelia Sheer
Location: Nashville, Tennessee
Date of Event: 2013
Witness: Michael (Assumed Name)

I'd like to tell you a story... a true story. Years ago, in doing my research, I came across a report in numerous places. Supposedly in the 50s, an individual with high government standing, with secret clearance on a black budget committee, was brought into the UFO world in a very sudden manner. In attending a budget meeting on a fateful day, he noticed that there were armed guards at all entry points, and security was extremely high. After settling in for a long meeting, he noticed three gentlemen of striking looks and impeccable dress standing in an area where all could be observed. He thought that was a bit unusual, but this was Washington, and strange things were the norm here.

As the meeting was called into session, there was an immediate change in protocol, an announcement was made about the seriousness of the day's agenda, and non-disclosures would be given to each individual to sign. The terms of those non-disclosures were explained in detail, and the severity of any kind of breach was reinforced again and again. The group was informed that the U.S. Government had made contact with an alien race and entered into a trade agreement with them. This race had agreed to share certain restricted aspects of its technology in a collaborative work with the U.S. in pushing forward our defense abilities. It seemed that they were aware of other races that were not so benevolent and were concerned about our inability to protect ourselves in any competent fashion.

To say the least, this government official was stunned! It was the 50s, and in those days, talk of aliens and UFOs were minimal, and any such claim received much ridicule. He was able to ask a few questions about the "aliens", and he learned that they looked pretty much like us and could pass easily walking down the street. Interestingly, he also learned that they were highly telepathic, so no attempt at deception would work or be tolerated. After coming to terms with all of this "shocking" information, he settled into the day's agenda, but his mind was working furiously. He began to suspect that the impeccably-dressed men were, in actuality, some of the "aliens", and he could not help openly staring at them and wondering about so many things. He began to form an idea that if they were really telepathic, he would perform his own experiment and send them a message to see if he would get any response. Of course, he had no idea how to be "telepathic", so he just doubled up and concentrated as hard as he could while focusing on the three gentlemen in front of the room. In his mind, he asked the men that if they could really hear him, to meet him at lunch out front on an open bench, to sit down and touch their nose. The time 'til lunch recess slowed to a crawl, but eventually the break came, and he hurried out to an open bench and waited. Sure enough, one of the striking, slightly tanned men came out, sat down, touched his nose, and smiled the biggest smile at the man...the rest is history! This story fascinated me and captured my curiosity. So, me being me, I proceeded to send out what I thought to be a telepathic message anytime I was out in public. I've been doing this for 25+ years. Restaurants, airports, bookstores (my favorite)...it didn't matter, I would always take the time to silently call out and hope that one day I might get a response.

Years rolled by and nothing...then one day about 6 years ago, I was in one of my favorite cafes, got settled, bowed my head (most other patrons thinking I'm praying), and sent out my little telepathic message. "I'm Angelia, I have blond hair, and I'm sitting alone, reading a book. If you hear me, please come talk with me." At the moment that I completed my petition, I looked up, and a striking man (I'll call him Michael) got up from his table, walked directly to me, and, with a twinkle in his eye, asked, "Can I have lunch with you today?"

Needless to say, I was stunned. The first thing he asked me was, "What ya reading?" I told him it was a book on UFOs (of course), and he said, "Now we can have a fun conversation!" I thought my heart was going to beat out of my chest, and the strangeness did not stop there but continued on for the remainder of our lunch. At first, I was thinking, "Wow, I just want to get this guy's picture to validate he's really here." I was also hesitant to ask as I had just met this man, and it seemed rude. At that moment, he placed his hand on mine and asked politely, "Can I get your picture?" The remainder of the lunch followed that same pattern. Whatever I thought, he would say in the next sentence or so. It was astonishing and beat the odds on any kind of coincidence as this did not happen just a few times but went on the entire 1-1/2 hour lunch.

On parting, he said to me, "I want you to know that I'm just a regular guy, so I'm going to buy your lunch, and then I want you to walk out to my car with me and meet my dog." We strolled out to the car, and I was introduced to Sasha, a big bundle of love. We talked a few more minutes as he told me how much Sasha meant to him, then we exchanged numbers and got into our cars. As I drove away, I was thinking it was a bit warm today and I hoped the dog had not gotten too hot in the car as we shared lunch. In that moment, my cell rang, and it was Michael. He stressed to me how much he loved that dog and how he would never let her get too hot, that her well-being was one of his major concerns. I just drove back to work with my mouth hanging open. Had I just met a real, live "alien"? I had searched the world over for 35+ years for individuals who reported UFO sightings, high strange events, or who seemed to possess unusual abilities, and here I had just sat, eating lunch with a handsome man who seemed to read every thought I had for almost 2 hours!

Several weeks went by while I secretly wished that I really had met an alien. I'm very non-dramatic, don't embellish, and was sure of what I had experienced that day. It's like hitting the lottery, and you're just waiting for someone to take it away. I wanted to call Michael and question him some more, but in a way, I was afraid to find out that he wasn't an "alien", and it was just a fluke. These thoughts still haunted me, and one day while driving again to that same café, my phone rang. It was Michael, and he said, "I thought you would be here today." (Meaning that he was already at the café and somehow knew I would be there, and here I was, driving to the café.) I'm still stunned to this day, and I promise to all of my brave friends out there that this is a true story with no embellishment. As time went by, I revealed my passion about aliens and such to Michael and asked if he would discuss his abilities with me. I also asked him point blank if he was an "alien". He smiled at me with that twinkle in his eyes and shook his head no. But something deep inside me still wonders!

To this day, Michael and I are friends and meet at times (on his schedule) to talk and just hang out. Due to his sensitivities and abilities, he is much of a recluse, and sometimes months will pass with no communications. He is extremely intelligent, has authored numerous books, is emotionally sensitive, is a talented musician, and to this day, still displays these moments of incredible telepathy. I see his struggle in interacting with the general public and have experienced his utter fatigue and, in some sense, confusion in dealing with what appears at times to be a complete sensory overload. I wonder how many others there may be out there who also have these same abilities but stay hidden away from the onslaught of unchecked emotional chaos that seems so rampant in our society today.

As I write this --imagine that! -- Michael has contacted me, and we are planning to meet for dinner one evening. He says he has much to catch me up on and, when asked, agreed for me to share his story as long as I keep him and his life confidential. This is a true story, and its implications for humanity are huge and far-reaching. I hope my friendship with "Michael" takes us long into old age and that great adventures still await us. - UFOgirl

His Response: A Message from Michael >>>"Human life is complex and challenging, and I don't want to make it more so by misleading anyone. Everything happened just as Ang described it, but I was not consciously aware of reading her mind that day, though it seems I probably did. I have done that with other people from time to time. And, though I have had more than my share of paranormal/psychic experiences, I am not usually able to manifest them at will. I think of it as a little helpful edge that I have been given for when things are critical. Yes, I have been told that I am not a human being, but all of us are just spirits having human experiences. I believe that the paranormal and miraculous surrounds us all of the time, just like God's love!

Case: Going Up, Please
Parasheer Research Files
Investigator: Angelia Sheer
Location: Knoxville, Tennessee
Date: 1984
Witness: Sgt. Randy Cutshaw
Bio: Deputy Sheriff/School Resource Officer going on 19 years, Officer of the Year in 2011, School
Resource Officer of the Year in 2013, married for 21 years

Author's Note: I did a search and found numerous reports on Sgt. Cutshaw's many arrests and achievements.

As I was wrapping up writing this book and thought I had all the stories that I wanted to share included, I received a very intriguing message. The email was from a Sgt. Randy Cutshaw, and he had a very interesting story to share with me. I have had many pilots and police officers share strange sightings with me, but due to the stigma involved in reporting such things, most want to remain anonymous. I did some checking on Sgt. Cutshaw's background and found many articles and reports concerning cases and arrests that he was responsible for, so I was excited to speak with such a credible witness.

After a few missed calls back and forth, we finally connected, and only after a brief time, we were talking like long lost friends. I was again reminded how strange the UFO/Paranormal mystery can be. As I have intimated along the way, witnesses usually fall into one of two camps, either they have had sightings/high strange events all of their lives or they have a sighting later in life that seems to become a defining moment in their growth and perceptual abilities. Randy falls into the first category with the report of something that happened to him at a young age. Pay attention to the details of his story and how, as he matured, he continued to experience Paranormal Events as well as a pretty dramatic UFO sighting with his daughter. So, in his words, here is Randy's first defining experience…

"The summer of 1984 became the most unforgettable day of my life. I was 15 years old and had not a care in the world. The age of no cell phones, iPads, internet, video games made us play and entertain ourselves outside. Every summer during school break, me and other friends would ride our bicycles down to DouglasLlake off of Indian Creek Road. This particular day was me, my sister, my uncle, and my neighbor all near the same age. Not one of us could swim so we would just get out far enough, but not to where it was too deep. Earlier that

day, I found a life jacket that had seen some better days, but it did allow me to float around some. We made it to our same spot at the lake, and I put the life jacket on. I was floating in the lake for around 10 minutes when my uncle said that I was floating out too far. I immediately panicked, and for some unknown reason, I took the life jacket off. I sunk straight to the bottom of the lake, probably around 10 feet deep. While under water, I remember someone saying a few weeks ago that if you go under water that you can wave your arms up and down to get back up. I began flapping my arms up and down without any success of getting air. At one point, I could see everyone on the bank yelling and screaming but was unable to help. After about 1 minute or more of trying to get air, I remember thinking that I was going to drown in the lake. I was not scared, terrified, or sad. I was just going to accept it and die. As I was about to give up and drown, _a solid object came up underneath me_ and started lifting me up, but with flapping of the arms and legs, I slipped off and sunk to the bottom again. The solid object came back from underneath again, but this time, I stopped flailing my arms and legs and let it lift me. The object lifted me waist high out of the water, and I was using my hands to wipe the water from my face and trying to get fresh air in my lungs. This object felt like a solid piece of concrete floor, it did not float or bob in the water. Whatever it was, it was very strong. When I opened my eyes, I noticed that my neighbor had a long tree limb for me to reach out and grab. I grabbed the tree limb, and they pulled me to the bank, but with all the chaos, no one noticed what was really going on. Everyone was exhausted from being panicked after watching me almost drown, and we sat down on the ground trying to rest and breathe. When we had all settled down a bit, I told them what happened to me under the water. Thirty-five years later, and to this day, we have never spoken again about the day I almost drowned. - Sgt. Randy Cutshaw

Photo by Randy Cutshaw

Just to the right of the small tree is where the witness almost drowned. The witness shared that the water level was way down when he took this picture.

Here is a classic example of a paranormal experience that actually saved someone's life and at the same time conveyed an experience so powerful, the existence of high strange events became a reality and not just something to speculate about. I shared with Randy that, at about the same age, I experienced a similar life-changing event, so I never had to be convinced about the existence of paranormal intrusions into our world...I _knew_ they existed. Also, pay attention to the fact that he and the others all experienced what we define as a "high strange event" and yet these experiences are in most cases minimally discussed and eerily dismissed. And, as our conversation unfolded, it became apparent that Sgt. Randy Cutshaw had experienced these strange encounters many times in his life…

"December 28, 2011 was a very sad day for my family and me. My father in-law passed away after a lengthy illness. On the morning of the third day of his death, my wife was upstairs getting ready for work when she heard three knocks coming from somewhere inside our house. I didn't hear it, but she thought my daughter or I was playing a joke on her. The very next morning around the same time, my daughter heard three knocks from an unknown location from inside the house. Again, I didn't hear the three knocks, and at this point, my wife and daughter are thinking that I'm playing jokes on them. We were at the dinner table talking about the knocks, and I stated that I wanted to hear the knocks. The very next morning, which was the third morning, I heard three distinct knocks very loud and clear. I thought my wife and daughter were playing the joke on me, but when I started to investigate the knocks and went to the upstairs bathroom where they were, they had no clue of anything going on because my wife was taking care of my daughter. I silently stood outside the bathroom door to see if I could hear some sort of clue or to see if they were laughing. When they got finished in the bathroom and came downstairs, I told them what happened. They had no idea what was going on. The knocks stopped after the third day."- Sgt. Randy Cutshaw

Here our UFO/Paranormal connection continues with the intrusion of unidentified knocks in the house. The onset of these phenomena is very common in the aftermath of many UFO sightings and is reminiscent of an account described by the great Whitley Strieber in his 1988 book, *Transformation*, pg. 130, *"At that moment there came a knocking on the side of the house. This was a substantial noise, very regular and sharp. The knocks were so exactly spaced that they sounded like they were being produced by a machine. Both cats were riveted with terror. They stared at the wall. The knocks went on, nine of them in three groups of three, followed by a tenth lighter double-knock that communicated an impression of finality."*

Notice the pattern of 3 knocks in succession. Why 3 and why knocks? Notice the similarity to Sgt. Cutshaw's description of what he experienced and his belief that there was no natural or simple explanation for those knocks. And, these events are being reported by a trained police officer who is certifiably experienced in investigations, calm, not prone to embellishments, and well respected in his community. Again, I say, "You just can't make this stuff up!" After at least 2,500 interviews, I am still amazed at the UFO/Paranormal pattern that emerges again and again in credible witness testimony. As Sgt. Cutshaw and I continued our conversation, he lamented that he had only had one UFO encounter, but he could go on and on about all of the high strange events he had experienced. As our pattern plays out again as documented with thousands of other accounts, our witness continued to experience strange events until finally in 2009, the unknown came knocking again with this encounter...

"In August of 2009 around 7:00 a.m., my daughter and I walked out of the house and started to go to work, and I was going to take her to school and drop her off. I approached the driver's side door of my patrol car and opened it. As I began to get in the car, I saw an object in the sky that was a pinkish purple and green color in the corner of my eye and was traveling at supersonic speed with absolutely no sound at all. I sat in the driver's seat, and my daughter, who was 9 years old at the time, verified that she also saw the object. I asked her what she thought it was, but she had no idea. I watched the object for only a short time as it was really moving then proceeded to get back out of my patrol car to take a better look but was only able to see what remained of a smoke trail left in the sky. We were both baffled by that object, and to this day, that sighting remains a mystery!" - Sgt. Randy Cutshaw

I cannot express how much fun it was speaking with this witness! He is not only a hero in my eyes, risking his life each day to protect his community, but also brave enough to share some incredible events he has experienced over his life. My special people…my friends…have really looked through windows to other worlds, have gathered courage to explore these mysteries, and have come back to us, sometimes at great risk, to share their accounts. Many times, I compare this situation to our current legal system which is predicated on credible witness testimony. This testimony can put you in jail and sometimes put you to death. Why would sane, stable people fabricate and share events that are in no way actually helpful to their lives. The events that I have chosen to share come from that kind of credible witness testimony. We should listen and listen carefully! Just as sages, shamans, and seers of the past invited their people to go beyond the known, our witnesses of today also offer an invitation to go beyond our rigid ideas of reality, to step into the unknown and possibly begin a transformation in the process. Thank you, Sgt. Randy Cutshaw, for another act of bravery in sharing your stories, and I hope to share in more of your adventures in the future! – *UFOgirl*

Author's Notes:

1) Witness experienced a very high strange event as a child that saved his life.
2) After that event, the witness experienced UFO Sightings and a mixture of other paranormal events.

Case: Strange Lights in the Bedroom
Parasheer Research Case Files
Investigator: Angelia Sheer
Location: Nashville, Tennessee
Date of Event: March 2016
Witnesses: Dale Houston and Angelia Sheer

I love my work! Anyone who is around me, even for a short period of time, will realize that immediately, as I am incapable of not talking about it. Dale Houston says that it takes less than 9 minutes in any situation before the topic of "UFOs" finds its way into the conversation. Other worlds, other beings, endless adventures, how could you not be curious about that? Over the years, I have been fortunate in the cases that I have had the honor to investigate, so many mysteries to unravel, places to explore, and people to talk to. So, when the "paranormal" comes knocking on my personal door, that turns things around a bit.

To set up the circumstances surrounding my strange event, I have to take you back and give you some background on my partner in crime, Dale Houston. Many years ago, when Dale and I were first together, I began to notice a pattern of behavior that revealed itself as Dale slept. Many times in the middle of the night, Dale would sit up in bed, eyes open, seemingly completely awake, and start talking to me about really strange things. One of my first experiences with this "altered state" was a late-night conversation early in the relationship that clued me into what was happening with Dale during these events. It was around 3:30 a.m. (make note of the time) when all of a sudden, he sits up in bed and starts talking to me. Being a mom and being infinitely sensitive to the ones I love, I immediately came awake as I thought something might be wrong. I turned over and Dale was sitting up, eyes wide open, and he just starts talking to me. He reached over tenderly, brushed my hair back, and in quite a romantic tone told me the following. He said, "I know you have been hurt in the past, and I promise I will never do that to you. You don't have to worry, because I've gone on ahead and checked things out, and it's all going to be okay!" Well, being that our relationship was in the early stages, I was quite moved by this tender display of affection and concern, told him so, and eventually we both drifted back off to sleep.

The next morning, I was down in the kitchen making breakfast (per Dale, as a good woman should) when Mr. Houston finally smelled the bacon cooking and made his morning appearance. I was happy and excited about our conversation the night before and immediately thanked him for his kind words. I will never forget this as long as I live. Dale looked at me straight in the eyes in a confused manner and said, "I don't know what you're talking about," and he really didn't. I was shocked! I could not believe that he could seem so awake, so cogent, interact so tenderly, and then only a few hours later have no memory whatsoever of that conversation. It was in that moment my training in meditation, altered states, and the extensive training I received at the Monroe Institute came flooding back to me. It became clear to me that in these late-night conversations, Dale was experiencing an altered state and was quite possibly tuning into other areas of information, so the researcher in me went to work. From that day forward, whenever Dale woke me up in the wee hours, I was ready with notebooks and a plan to question him about what he was experiencing.

What has transpired over the years since that first late-night encounter would have most of you rolling around on the floor laughing. The human psyche is unbelievably complex, is built for survival, is wrapped in layers of defenses, spans across an impressive range of the electromagnetic spectrum, and it amazes me to this day. Many nights, Dale would spring awake, get up, and start getting dressed. He would grab his phone, put on his pants, grab his boots, and rush around the room, seemingly fully awake. On one of these first encounters, I asked him what he was doing, and he responded, "I have bus call, and I have to go." Dale grew up with "Merle" on a tour bus and, to this day, still has "Ramblin' Fever" as he spends much of the year traveling and singing across the U.S. Of course, there really was no bus call that night, but he was convinced that there was, and no amount of my arguing with him was going to change his mind. He actually would get quite put out with me trying to convince him that he was sleeping and that he really didn't have to go anywhere. It took some time for me to figure out how to deal with this situation as trying to convince him in this altered state was just futile. Eventually, I would not argue at all, I would just say okay, express how lonely and cold this big empty bed was, pull back the covers (you guys know what I'm talking about), and he would immediately haul his big butt right back into the bed with much enthusiasm. I'm cracking up laughing as I write this, I giggle to myself every time it happens, and it continues to this day! A note to all of my young aspiring researchers out there: this is an example of what we call "boots on the ground" research where sometimes there are no guidelines to follow, "punting"! Furthermore, as time went by and our late-night research continued, Dale many times when I was questioning him in this altered state would actually tell me that he was not sleeping without me ever trying to convince him of that. That's how shrewd our consciousness and beings really are!

So, as time went by, I got to study Mr. Houston's nighttime adventures up close and personal. I experimented in talking with this "other" personality (a little reminiscent of people like Edgar Cayce who slip into altered states and can deliver quite elegantly what they are perceiving), and as time passed and my ability to interact with this other part of Dale improved, he would answer my ongoing questions with sometimes exciting results. One early morning, around 3:00 a.m., Dale sits up and starts talking. I'm immediately awake and ready to question him about what's going on. As I turned over, Dale is turned away from me, looking at the area beside his side of the bed. I can tell he is scanning the area and appears to be actually seeing something. I asked him, "Hey baby, what are you seeing?" He immediately responds, "They're building something over here, and I don't like it." I asked him to describe what he was seeing, but he seemed to be having trouble finding something to compare it to. It was like he was seeing something for the first time, and he had no context to fit it into. He continued to look around the area, sometimes actually appearing to be looking out into some distance even though the wall on his side of the bed is only a few feet away. I made note of everything I could get out of him about what was happening before he became too frustrated in trying to relay what he perceived. Of course, in the morning he had no memory whatsoever of the event.

Immediately, this particular late-night adventure piqued my curiosity as it was really quite different than most of the others, and I couldn't stop thinking about what Dale might have been seeing. No amount of questioning while Dale was awake brought forth any new information, and I could not stop pondering on how I could maybe discover more about what was reportedly "being built by the bed". About two weeks went by, and one night I got the idea to video our bedroom in the dark. Don't ask me where this came from, but these spontaneous feelings and ideas seem to have merit as many of my witnesses have shared with me. I called Dale into my office and told him that I wanted him to get his new phone which had a really great camera and wanted to make a video in the bedroom. Well, all of you who know Dale know what his response was to that! Never a serious moment…so after dashing all of those ideas, we both headed up to the bedroom to start our experiment.

It was around 9:00 p.m. that we finally got settled and got the video running. We had all the lights turned off and just began panning the camera around the room while lying on the bed. Shortly, we started seeing what appeared to be tiny flashing white lights all around the bed. At one point, you hear an intake of breath and then hear me exclaim, "What is that?" The lights continued dancing around the bedroom, and we kept filming. We could not see the lights with the naked eye, but they were clear as day on the screen of the phone as we were recording. We continued recording for about 2 minutes, and as the phenomena faded, we ended the recording. After that, we watched that video I don't know how many times over, and we also tried numerous times to reproduce it. All in all, we made six other attempts to try and create the same flashing lights, but we never could. It seemed that we really had captured something strange happening in our room. Furthermore, we discovered that around the 7-second time stamp, an audio phenomenon can be heard. We did not hear this while we were recording. It stops and begins again around the 40-second mark and continues until about the 1-minute mark. The audio capture sounds somewhat like a channel is changed on the radio and you hear what could possibly be something talking. After much scrutiny, it seems that two voices are heard, with one being closer and one off at a distance. This was really surprising for both of us, as again, it seemed that we really captured something paranormal in relation to Dale's late-night adventures.

What happened next is completely mystifying. The next morning, I was again downstairs cooking breakfast when I heard Dale come to the top of the steps and call for me to come upstairs. I could tell he was agitated, which is strange enough for a chilled-out musician, so I hurried up the stairs and into the bedroom. He was actually in our walk-in closet and called me to come in and look at the video we had taken last night. I asked him, "Why?", but he didn't answer, and he just ushered me into the closet. The reason he was in the closet is that our house is full of windows and that's about the only place that's dark enough during the day to see a video made in the dark. He at once started playing back the video, and after my eyes adjusted, I started watching. Immediately, I was aware that the white flashing lights we captured the night before were gone. I told him that he must have the wrong video pulled up, but he stopped me and said, "Listen, you can hear the audio distortion at the 7- and 40-second time stamps," and you could. I actually had a momentary mental shutdown. It just couldn't be possible that both of us could have seen those flashing lights last night on that video, and now they were just gone. But, sure enough, we watched that video over and over standing in our dark closet, and we could hear the audio distortion come in, but those flashing lights were gone. We were both freaked out a bit and completely mystified. There wasn't much else to do at that point, so we put the video away and just went about our day.

Later on that evening, a friend who is a complete skeptic about these matters came over to dinner, so Dale pulled up the video to at least let me hear the audio distortion we had captured the previous evening. And, guess what, the lights were back, I kid you not. To this day, even though I have dealt with high strange events all my life, I'm not sure I would have believed this if I hadn't seen it with my own eyes, and yet there it was. It was like something just wanted to let us know that it could play with our technology like a child's toy if it so chose.

Case Updates: Sometime after this event, Deb White Kauble, AKA Kathie Davis from Budd Hopkins' book, "Intruders, The Incredible Visitations at Copley Woods", came to stay with us for a visit. While we she was here, we of course talked about current research and theories. One evening, Dale brought out our video with the strange blinking lights and weird audio, and Deb immediately commented that the audio sounded eerily similar to what was captured in the "Strange Case of Witness J", and she was right. Furthermore, "J's" case was reported in April of 2016, and Dale and I captured our video in May of 2016. Until that night, Dale and I had not made that connection, and to this day, other strange events continue to unfold here as well as with the case of "Witness J".

Photo Courtesy of Dale Houston

8. PERCEPTION "101"

"It is one of the commonest of mistakes to consider that the limit of our power of perception is also the limit of all there is to perceive."
                                                        - C. W. Leadbeater

"The most important thing is to not stop questioning. Curiosity has its own reason for existing."   - Albert Einstein

Now...before we can go any further, I must bend your mind a bit. This might get a bit complex, but I promise it's worth your time and attention. So far, we have examined some intriguing UFO cases and other high strange events. Each of the cases chosen came from credible witnesses, and they were "sure" in what they experienced and perceived. Think back to the case I shared with you in the introduction of the book that really changed the way in which I studied the UFO phenomena. I will refresh the reader's memory with a short recap here.

Case: The Beach Party

Years ago, a report came to me that changed how I viewed the whole UFO mystery. There was a group of about 14 individuals having a beach party one evening. Around 10:30 p.m. or so, 4 of the individuals started seeing a UFO out over the ocean. At the time of the sighting, all of the people present were within approximately 25 feet of each other, so I could rule out any type of relative viewpoint issue. The 4 witnesses starting yelling to all the other partygoers to look at the stunning sight that they were seeing. But, no matter how hard they pointed, yelled, and described the object, the other 10 people could NOT see the UFO. When separated, the 4 witnesses described the same object, colors, sounds, etc., and they were sure of what they had witnessed. Yet, the others saw and heard nothing. Many of my friends in the psychology field brushed it aside as drugs or some kind of joint hallucination, but I wasn't buying it! I have a 30-year medical background including nursing experience. Not once have I come across a group of people that had that kind of shared hallucination. I was sure that no one was on drugs or lying, so what actually happened? The 4 individuals who saw the UFO were sure of what they saw, and the other 10 individuals were sure that they saw nothing. So, what in the world was going on? After investigating the account, I was convinced that the 4 witnesses really did see something, and for some reason, the other 10 just could not perceive it. That understanding was huge, and at that moment, it changed how I viewed and then researched the UFO phenomena. For I was convinced that those 4 individuals could actually _perceive_ a greater degree of the electromagnetic spectrum than the others. And, with this greater degree of awareness, they were privy to a greater world of beings and interactions that surround us all the time. At this point, my research took a radical turn. From then on, not only did I research the external phenomena, but I began an in-depth look at the human component of each sighting. That's when things got really interesting because so much of what I had been taught and/or assumed about the UFO phenomena began to take on a whole new set of criteria. And, those criteria were a pivotal component in the understanding of the overall phenomena that just could not be disregarded. The one missing component to our study of the UFO mystery seemed to be the human being itself and its varying degrees of perceptual abilities. When this component was added to the study, a lot of things began to make sense, and 38 years later, my research holds. My witnesses who displayed a greater acuity in perception had the most sightings, and it all had to do with the development of their perceptions.

Human perception, now that's a tricky proposition. To truly understand the implications of our part in the creation of how we perceive what's out there, we must have an understanding of how human perception really works. And, for me to get you there, we have to start at the beginning. So, first of all, let's take a look at how our eyes create the things we think we really see. Most people believe that our vision is just like a camera, that from moment to moment, we are just taking a picture of what is really out there. That could not be farther from the truth!

In a general web search, we find:

*"The **images** we **see** are made up of light reflected from the objects we look at. This light enters the **eye** through the cornea, which acts like a window at the front of the eye...Because the front part of the **eye** is curved, it bends the light, creating an upside-down **image** on the retina."*

From there, the brain must take up the task of <u>interpreting</u> the <u>light patterns</u> that are displayed on the retina. That's where things get really interesting. The brain then must reach into its repository of knowledge to date and construct a facsimile of what you are looking at. Really think about the implications of how, we as human beings, construct our reality and what this means. To a huge degree, we are constructing and reconstructing our reality moment to moment. From the time we enter this world and take our first peeks "out there", we are indoctrinated into what it all means. This is a tree, that's a tomato, UFOs don't exist, other worlds don't exist, etc. So, to a huge extent, our early programming determines the world we perceive because it builds the library of choices that our brains have to choose from to construct what we see. Many of my witnesses who I have come to study over the years seemed to have built in sensitivities or developed said sensitivities, so they were able to perceive beyond the accepted standard and reach into greater worlds. As mentioned before, this skill set was not usually met with acceptance from family and community, so many individuals went underground with their abilities or just kept it all to themselves. This, I feel, is one of the greatest tragedies humankind has suffered, for these gifted individuals are our herald into the future of greater worlds and possibilities for humanity. But, I digress, so back to the problem of perception...

The understanding of what we perceive out there in relation to our consciousness is not a new problem and has been debated endlessly.

"I think that tastes, odors, colors, and so on...reside in consciousness. Hence, if the living creature were removed, all these qualities would be...annihilated."   - *Galileo*

This is an interesting interpretation but probably not the whole picture. I guarantee you that if you stand in front of an oncoming train and close your eyes, shortly you will be with us no longer. And, if I as a witness, were watching this event, if I closed my eyes at the moment of impact, you as my friend would still be no more when I opened my eyes again. So, where does the truth lie? Many in the New Age world tout that you create your reality each and every moment, but what I think they fail to remember is that we are creating that world in conjunction with billions of other humans. Sure, interpretations of events can vary radically, but in the greater scheme of things, we all come to at least a workable interpretation for day to day interactions. So, what's really going on? Probably some combination of both. We do have some power over what we create each and every moment, but that interpretation is constantly affected by the other billions of souls that live here with us, especially those that are closest to us.

I'm really fascinated by the work of Dr. Donald Hoffman. After much study and experimentation with human perception, he has concluded so far:

*"It is well established that perception is not a reliable copy of the external world but only part of it composed by external stimuli, while the rest is constructed by the **brain**. This means that the **brain creates** only the **reality** it is interested in for the survival of the organism."*

"In his research to uncover the underlying secrets of human perception, Donald Hoffman has discovered important clues pointing to the subjective nature of reality.

Rather than as a set of absolute physical principles, reality is best understood as a set of phenomena our brain constructs to guide our behavior. To put it simply: we actively create everything we see, and there is no aspect of reality that does not depend on consciousness."

Dr. Hoffman's work cannot be completely summarized here, but it can give you at least a basic understanding of the limits of our perceptual abilities and the interpretations we give them. For a more in depth look at Dr. Hoffman's work, check out his Ted Talk at:

https://www.ted.com/talks/donald_hoffman_do_we_see_reality_a

s_it_is

In summary, the basic facts of human perception can be reduced to this equation:

1) Light bounces off some external object, enters the human eye, and constructs a reflection on the retina.

2) The human brain then reaches into its current depository of experiences and learning and creates a facsimile of the external object.

3) There is not one act of perception that is exempt from this process.

4) The greater amount of accumulated knowledge and experiences an individual integrates has a direct relation upon the creation of not only the images we create but also their interpretation.

Conclusions from my Research:

1) Certain human beings display perceptual characteristics that exceed the standard norm.
2) These advanced abilities are existent at birth, develop after the onset of traumatic events or incidents of prolonged stress, and/or are developed through training. (i.e., The Monroe Institute, see next chapter)
3) Humans that embody these advanced perceptual abilities perceive and interact with radically different worlds than the general public, are aware of greater fields of information, and possibly herald the advent of greater human perceptual evolution.

9. TYING IT ALL TOGETHER

"Science cannot solve the ultimate mystery of nature. And that is because, in the last analysis, we ourselves are a part of the mystery that we are trying to solve."　　　　　*- Max Planck*

"Where there is no vision the people perish."　　*- Jewish Proverb*

The Monroe Institute

In my mid-thirties, I came across some incredible books written by the late Robert Monroe. At that time, I was studying everything I could get my hands on about human consciousness and artifacts of state changes. I had been lucky to have had lucid as well as out of body dreams from childhood that were exciting and mind-bending. I had actually been able to travel with other individuals as well as visit my friends while they were sleeping and then was able to document what I saw and compare notes. It was amazing that, statistically, many of us could describe rooms, books, beds, pictures, etc. while in these altered states with some degree of accuracy. Needless to say, this was amazing and flew in the face of what we understood about human consciousness and abilities. Granted, these altered states were well-known in ancient and esoteric traditions but were just pretty much ignored by modern science. And yet these events just kept on happening to me on a regular basis, so as any good child of a scientist does, I started keeping detailed journals of my adventures. I kept track of such things as diet, time of year, time of day, emotional states, physical states, and psychological trends and how these affected the altered states that I was experiencing. When I came across Robert Monroe's "Journeys Out of the Body", I was hooked and dug in and read everything he had. For more information, please check out:

For more information about The Monroe Institute, see: https://www.monroeinstitute.org/

Books by Robert Monroe: Journeys Out of the Body
Far Journeys
Ultimate Journey

Needless to say, I was impressed with Robert Monroe's credibility, excellent work, and science. I immediately applied to attend courses and was accepted as a student for my first "Gateway" program in 1996. I cannot stress enough how this Institute nestled in the Virginia Mountains changed my life forever.

First of all, I was surrounded by a small group of individuals from around the world with many of the same talents and interests. There were doctors, physicists, housewives, and every kind of individual represented within our group. We all had shared to some degree some pretty amazing events and were determined to understand and possibly learn how to reproduce some of those events. We not only practiced altered state maneuvers together but spent endless hours discussing our experiences. All in all, I attended 11 courses spanning over 14 years of training with some incredible facilitators and students alike. Below are some of the Programs that I attended:

Gateway	1996
Guidelines	1997
Lifelines	1998
Exploration 27	1999
Lifelines	2000
MC2	2001
Timelines	2003
Starlines	2004
Starlines	2009
Starlines II	2010
Excursion Workshop	2010

I cannot go into all of the events and details of what transpired through those years, but I can tell you that I experienced incredible out-of-body states, witnessed telekinetic effects, and personally experienced numerous telepathic encounters that were verified with other members of the group. *I was firmly convinced, first hand, that humanity is equipped with many astonishing abilities that are part of our being, can be trained, and unfortunately are not readily known or acknowledged by our current scientific paradigms.* From a young age, I was a disciplined meditator, and I saw the depth, quality, and integration of these practices deepen exponentially with every program I attended at the Monroe Institute. It was amazing what happened over time. My sensitivities, learning curves, study retention, and altered state activity went off the charts. It changed my life for the better, and this Institute, its staff, and fellow students will forever hold a precious place in my heart and soul!

You may ask, what in the world does altered state training have to do with the UFO/Paranormal phenomena? That's a great question...and the answer is "everything". So, to bring you up to speed, I want to run through just what we did at the Monroe

Institute and the brilliant insights Robert Monroe had about his experiences. Remember, Mr. Monroe had extensive experience in the broadcasting industry, and this played an important part in helping him investigate what was happening to him. The following is taken from the Institute's website, and I encourage all to visit the site for more information about this technology, its development, and programs offered.

From the Monroe Institute's website:

"While continuing his successful broadcasting activities, Monroe began to experiment with and research the expanded forms of consciousness that he was experiencing. He chronicled his early explorations with a reporter's objectivity and eye for detail in a groundbreaking book, **Journeys Out of the Body**, *which was published in 1971. This public record of his out-of-body experiences in states beyond space, time, and death has comforted countless people who've encountered paranormal incidents. It also attracted the attention of academic researchers, medical practitioners, engineers, and other professionals.*

Ever the pragmatic business leader, Monroe and a growing group of fellow researchers began to work on methods of inducing and controlling this and other forms of consciousness in a laboratory setting. This research led to the development of a noninvasive and easy-to-use enhanced binaural beat audio guidance technology known as hemispheric synchronization or Hemi-Sync®. In 1974, the original research group was expanded to become The Monroe Institute, an organization dedicated to conducting seminars in the control and exploration of human consciousness. A year later, Monroe was issued the first of three patents for "Frequency Following Response" or FFR, which is part of the Hemi-Sync method of altering brain states through sound."

In the programs I attended at the Institute, the exercises were built around the Hemi-Sync technology. We would lie comfortably in our "CHEC Units", (Controlled Holistic Environmental Chamber), put on headphones, and then be sent the Hemi-Sync exercises throughout the day through multiple daily exercises over a 5-day course period. And, wow, what we experienced, shared, and documented was truly incredible. Many individuals shared similar visual and auditory experiences as well as some truly incredible telepathic events. Also, people were reporting communication with different types of entities as well as reporting seeing UFOs and other paranormal events. And this is while being isolated in a mountain retreat in our individual CHEC units. Monroe was truly a brilliant man and observer of what was happening to him and to others. He created a system of learning and common language that enabled his students to not only experience altered states from his Hemi-Sync technology but to also create personal reference points that would allow his students to experience said states without the technology. We all just kept being blown away with what was happening.

Here I want to define a few things so this will be very clear to my readers as this is very important to my research and conclusions. Some definitions:

Altered States: We all experience different states of consciousness every moment of our lives. From normal everyday awareness, to daydreaming, to deep sleep, to dreaming, hypnosis, and other forms of altered awareness including drug- and alcohol-induced states. Have you ever been driving down the road and you reach your destination and don't have any memory of how you got there? Well, that's a great example of an "altered state". You know you drove there but really have no memory of the journey. Kind of like missing time!

So, getting back to my training at the Institute, Monroe created a series of what he termed Focus Levels. They were numbered, for example: Focus 10, Focus 12, Focus 21, etc. Each exercise was created using certain combinations of back-engineered brainwave activity and were different for each Focus level exercise. What was incredible was that many of us were experiencing the same phenomena while listening to the same exercises. I remember one course called "Exploration 27" where well over 50% of the class described seeing the same place, same light, same lake, and same trees during the exercise. That's incredible and telling! As years passed and my training continued and the shared experiences just kept piling up, it became apparent that many of these places and/or entities we were describing to each other seemed to be real places and real entities. And this was happening while we were sequestered in our individual units by just changing the "internal channels we were aware of".

For example, every time I visited Focus 27, the same buildings and landscape would appear. When I would pass through Focus 12, a little gray alien would blink his eyes at me, and he seemed to be there every time I would visit Focus 12. Now that's really a game-changer! So, when I applied all of my altered state training to what was happening to my UFO witnesses, some very apparent patterns began to emerge. Here is where it really gets interesting!

Remember back to the end of each chapter in this book. There is a series of statements concerning the events that took place with each witness of that section. These points are actually indicators of events that cause state changes to happen. So, in every case reported to me, there are clear indicators that the witnesses experienced an environment where events transpired that caused their "state" to be altered. Events that can precipitate state changes can be almost anything, but a few common ones are:

Driving along dark, monotonous roads
Watching TV
Studying
Listening to music
Meditation
Relaxation
Intense emotional shocks
Traumatic events: There is so much to say about trauma and its effects on each of us. There was a great book written by Kenneth Ring, *"The Omega Project: Near-Death Experiences, UFO Encounters, and Mind at Large"*, that was a study in high strange events and its effects on the individuals experiencing them. Many skeptics in the past had tried to label individuals that reported UFO encounters and other paranormal events that had suffered trauma in their pasts as suffering from some kind of emotional pathology. What my research has shown, is that highly traumatized individuals do report a high number of extraordinary events but not because of their pathology but because trauma causes dissociation and dissociation is the key to awareness of other "channels" of awareness across the electromagnetic spectrum. I have often thought that schizophrenics are actually perceiving other realities, but the problem is that they have a damaged filtering system. Imagine being in a locked room with hundreds of televisions blaring away 24/7 and no way to turn them off. The perception of other realities is not what causes mental collapse but the mental overwhelm caused by constant over stimulation and the inability to turn it off. Many sensitive's that I have worked with and who function normally have always stressed the need to learn to temper their gifts, meaning, the ability to turn it off or on by choice. The saying, "The same water a mystic swims in is the same water a psychotic drowns in" is poetically very accurate!
Falling asleep
Multiple cycles of REM sleep (Isn't it interesting that many UFO sightings and high strange events happen around 3-3:30 a.m. Isn't

it also interesting that humans experience 90-minute cycles of REM sleep throughout the night. So if most people go to sleep around 9-10 p.m., by the time 3 a.m. arrives, they have experienced about 4-5 cycles of REM at which point you have reached some of the deepest sleep, achieved deep Delta Wave activity, and are experiencing some of the greatest altered states of the day.)

Drugs: Alcohol, hallucinogens, and even pain killers

Hypnosis

Technology: Such as Hemi-Sync, secret military weapons, and alien technology

Telepathy: Telepathic communications

…and this list goes on.

I know this can be a bit complex, but hang in there with me just a little longer, I promise it's worth it. To help explain what I feel is happening to many of my witnesses, I created a simplified model for clarity. Remember, this is an over-simplified explanation, and the details for a more complex explanation will not be shared in this work but saved for a later book.

Remember in my introduction I shared a case that changed the way I studied the UFO phenomena. It happened on a beach with about 14 people having a late-night cookout and gathering. At one point in the evening, all of the individuals were in pretty close proximity to each other around a fire when 4 of the people saw a UFO hovering out over the water. It was about 2 football fields out and was about 200 feet in altitude. A pretty close-range sighting, indeed! The remarkable thing, though, was that only the 4 individuals could see the craft, and no matter what they did to point it out, the others could not "see". Many of my professional colleagues wanted to write this off as a group delusion, but this just didn't sit well and didn't seem to be the case. It truly seemed that these few individuals' perceptual abilities were somehow different and allowed them to see into a greater reality (actually a greater degree of the electromagnetic spectrum) where the others could not. Now, whether this was due to some inherent difference on the part of the individuals themselves or was precipitated by some outside agency is a whole other discussion but still very relevant. This case and then others that came later convinced me that certain humans perceived very real things that others could not. And, if this were true and it seemed so as the years went by and the cases piled up, then besides studying the events, I also needed to take into consideration the witnesses themselves and how they played a part in the overall perception of the events that were happening to them.

Let's compare human perceptual ability to that of radio frequencies along the electromagnetic spectrum. It truly is all about frequency reception to some degree, and this is dependent on the equipment you are using. Like radios, TVs, and cell phones are all designed to receive different bandwidths along a certain frequency range. In this example, let's say most humans have the capacity to perceive channels 1-10, but most are really only consciously aware of channels 1-3. Then you can add in natural sensitives and/or psychics who may be able to sense up to channel 6, and up through channel 10 takes some radical training outside of the norm. Then imagine that on every channel, something different is happening. Also remember you're working on a frequency range, for example, if your favorite radio station is 102, many times you can hear your station on 101-103. The station may not be as clear on 101 or 103, but you still can hear it even though static may sometimes be present. Isn't it interesting to relate this channel drift to entities and/or craft that seemingly appear solid but then just seem to fade out? I wonder if many of these beings and their technology are always there but our ability to tune in is greater at times.

Let's say, for example, channels 1-3 encompass our everyday normal world, channel 4 may have apparitions or ghostly phenomena, on channel 5 grays might appear, channel 7 may be home to tall blond beings, and cloaked UFOs show up on channel 10. The combination of "channel" perception is probably vast and is only restricted by the perceptual capabilities of the individual at any given time. These capabilities are affected by personal abilities, interactions with the environment, and probably effects from outside agencies (i.e., advanced alien technology, advanced human technology, possible inter-dimensional and other exotic life forms). Also, although some humans may have advanced perceptual abilities along certain lines, this does not exclude the probability that alien life on "our" basic channel spectrum exists elsewhere in our galaxy and has now and in the past been interacting with mankind.

Also, in gathering data over the years on my witnesses and their experiences, other patterns began to emerge that is relevant to our "multiple channels explanation". I began to see that many individuals who had a general negative attitude, presented with addictions, displayed unresolved emotional issues, and generally presented with out of control lives filled with chaos reported the highest number of negative interactions with the phenomena. It is interesting to note that many of these individuals who were able to take command of their fears and made positive changes in their lives went on to experience incredible transformative events. As time went on, the negative interactions lessened, and more positive events were reported. *For example, it's as if our basic emotional nature sets the predominant channel we live on and perceive.*

Just as a symphony with all the instruments playing different melodies out of tune creates musical chaos, so does a life with emotional, physical, and mental fractures foster a basic incoherent frequency. Let's say this chaotic frequency sits on channel "4", for example. And, let's say some of the negative entities seem to inhabit channel 4 (again, because their basic frequency is also matched to channel "4"). When these individuals experience any state change, they automatically gravitate or drift to channel 4 (their basic frequency realm) where other entities seem to reside that inhabit that same frequency realm. Many of my witnesses who were able to move into these transformative stages were seemingly propelled onward to more balanced channels up the spectrum. As these changes were integrated within, their basic frequency channel changed, and they moved farther away from areas that contained more of the negative experiences. Close association does matter in the frequency realm. If you sleep outside in the woods, you have an exponentially greater chance of coming across snakes, ticks, and wild animals than you do in your cozy home.

At this point, I don't want my audience to make overgeneralizations! This oversimplified explanation does not imply that only "negative people" will have "negative interactions". You have to remember that humans are multifaceted beings. We have intelligence that runs along a multidimensional spectrum. This means we have physical, emotional, intellectual, and spiritual components that make up our whole being. If any one of these lines of development is off balance, we must make adjustments in order to grow to the next level of our potential. Have you ever met someone who was highly intelligent but displayed impaired emotional development? Or someone who was very emotionally sound but had impaired intellectual development? These are some examples of how we grow along multiple "lines" of being. For a brilliant detailed explanation of this, see Ken Wilber's work, "Integral Spirituality".

I have often heard the phrase "when bad things happen to good people"! Can you now see that we can be "good people" and still have aspects of our being that are out of balance and hold us back from moving forward? Sometimes we actually have done all we "know" to do to change, and it takes an event so out of our realm of comprehension to move us along to the next phase of our development. This experiential "kick in the butt" is our greatest challenge as a race and also our greatest hope. For it provides the sometimes-needed push to motivate us into examining parts of our being that may need a lot of introspection.

Also, I want to point out that sometimes things that appear "negative" are really the precursors of great opportunities for growth. How many of us when we're fat and happy go out of our way to actually dig in and learn something new? But the moment a UFO shows up hovering over our car, or small entities invade our bedroom, or we turn up with 3 hours of missing time, many of us suddenly take up the task of inquiry that leads to transformation. And that work, that inquiry, that struggle with the hidden and unknown is what propels us into those transformative events. It's as if this complex world is designed to kick our butts into growth whether we like it or not!

So, at this point let's look at what we have discussed so far. It seems all beings live along defined levels of awareness (channels/frequencies). The levels of awareness so far discussed are ones that are inherently present or have been achieved from work of the individual we are examining. Also, any changes on this channel spectrum are initiated by the individual consciously or unconsciously. Remember back to the list of circumstances that can cause state changes as these state changes are the precursors to channel changes. But what happens when an <u>outside agency</u> interacts with us to forcibly change the basic channels of our lives? Now this is where it gets really interesting!

Let's say we're just going along, minding our business, when something may come along to drag us to another channel. Now this kind of channel change can happen in either direction. We can be pulled down the frequency spectrum or pulled upwards. Kind of like changing the station on your radio dial. These channel changes can be precipitated in several ways by outside agencies. They could use technology (such as hemi sync, advanced military mind altering devices, or alien technology), drugs, or the actual "presence" of advanced humans or other entities could radically cause a state change through resonance or entrainment. What is resonance versus entrainment, you say?

<u>Sympathetic Resonance:</u> *applicable definition:* "*Sympathetic resonance or sympathetic vibration is a harmonic phenomenon wherein a formerly passive vibratory body responds to <u>external</u> vibrations to which it has a harmonic likeness.*" <u>Understanding **resonance** is essential for solving problems of increased vibration.</u>

<u>Entrainment:</u> *applicable definition:* "*a natural phenomenon in which one entity resonates synchronously with another in response to its <u>dominant frequency of vibration</u>.*"

Formula for state changes: or in our simplified model, "Channel Changes".

<u>Sympathetic Resonance</u> (changes that affect similar systems): These changes are experienced usually in a peaceful manner and lie along similar channel ranges. This is like meeting someone or listening to music, etc. that really "vibrates" very similarly to you! The channel changes are experienced as smooth transitions with little or no discordant properties. Remember this can be experienced from any channel transition and is reminiscent of the phrase, "Birds of a feather flock together". We always feel more comfortable with people who are similar to our own being.

1) <u>Relaxation</u>: You loosen the lock on the channel you are currently residing on.
2) <u>Dissociation</u>: You actually move off of the current channel and begin transition to a new channel. In sympathetic changes, this is experienced usually as smooth transitions.
3) <u>Reintegration</u>: You actually lock into the new channel and are able to experience information that is present in this frequency realm.

<u>Entrainment:</u> (changes that affect dissimilar systems): Such as being pulled upwards into higher unknown channels or even being pulled lower into more negative chaotic channel systems. These channel changes are usually experienced as discordant transitions. They usually cause energy drains, confusion, and some degree of anxiety. And this can happen in both directions. For example, when you are in a very negative environment, even if you are a positive person, you can hold out with your positive channel for a time, but eventually you will tire and be dragged down the channel spectrum. Similarly, when confronted with a higher yet unknown channel, you will experience eventual fatigue at holding on to your known channel (or current level of development or worldview) and be pulled, albeit temporarily, into the new channel. I have noticed that the level of fatigue experienced is directly related to our levels of resistance versus surrender to the impetus to move to a new channel.

1) <u>Relaxation</u>: Again, as in Sympathetic Resonance, you loosen the lock on the channel you are currently residing on.

2) <u>Dissociation/Disintegration</u>: You actually move off of the current channel and begin transition to a new channel. In entrainment, this is usually experienced as a discordant transition. I believe this is experienced in this way as you are actually experiencing a form of boundary disintegration. Whereas before this experience, your level of development only took you to channel "5", now your new awareness takes you to channel "6". I believe we all have inherent psychological/spiritual boundaries in place that are there for our protection. True growth is a tricky proposition, pushing us on to new levels of awareness. These transformations/transitions have to usher us into new realms of awareness but not too new or different as to render us too destabilized to deal with it. I used to see this with many of my meditation students who wanted to "hurry" their growth. "Speed kills" or "make haste slowly", I used to say. No change in our lives is equal to stagnation, too much change and you can so destabilize the current system of residence, you will introduce sometimes catastrophic circumstances that are debilitating physically, mentally, and emotionally. In essence, you have to build up enough psychic energy for a change or to actually break through the boundary that currently protects your level of awareness as well as restricts your movement upward (i.e., this is actually experienced as anxiety as the energy increases in an ever-confining channel system). Then when you finally can move to the next channel upward, you will probably experience relief as the energy can expand in a larger system. Like a plant has to be periodically repotted in ever-increasing larger pots. At this point in your new growth, you will probably not be able to maintain access to the new higher channels,

but you will disintegrate back down to the lower familiar channel until you can completely integrate the new levels of being. This is experienced in all areas of learning and growth. Remember when you learned to ride that first bike, and it took all of your concentration to just stay up? Now you can jump on that bike and ride away with little or no thought.

Partial Reintegration: You cannot completely lock into that new channel yet but experience fragmented moments of clarity on that new channel and then fall back down into known levels of awareness.

Final Reintegration: This is where you have seamlessly mastered movement along all your current "channels" and have the ability to change channels at will with precision. With this movement to a higher degree of complexity (integration of more complex information and ways of perceiving), you are forever changed and, in some sense, can never go back to the way you were before. Also, we must be careful in defining some channels as negative vs. positive. I have experienced very negative entities, as have my witnesses, as well as very positive beings. We must remember that "channel" growth is somewhat similar to going through school. We must master each grade before moving to the next level. We transcend and "include" all levels as we move up the scale of development. Each level becomes larger in its complexity of information and understanding, just as we must learn basic math before we can move to on to algebra, geometry, and then on to calculus. Then to complicate matters, there are good and bad teachers on every channel all along the way. Some overly soothe us into complacency whereas others kick our butts into growth.

Now go back and think about all those clues planted in the Author's Notes along each of my witness' reports. In every case, there were elements present that were indicative of major state changes either by the witness themselves, the environment, or the presence of seemingly alien entities. That means that either the witness themselves generated the state change that allowed them perception to other realms or they were affected by outside agencies that forced or aided in their transitions in perception to extraordinary awareness of other beings, ships, and/or realms. Let's also assume that many of these other "realms" are always there but are only limited in our ability to tune into them. This can explain a lot of things:

- We live in an incredible universe made up of ever-increasing levels of complexity.

- There are beings throughout our galaxy that are similar in their structure to us and that are aware of the same channels that we are aware of and exist on.

- There are probable beings that completely live outside of our galaxy and our known channel system. Some are aware of us, some are not.

- Advanced beings that exist outside of our known frequency realms (the channels that we are aware of) may have created technology that can not only interact with our worlds but can affect our ability to perceive and interact with their worlds, albeit sometimes only temporarily.

- And finally, when we interact with these other worlds and beings, we are fundamentally changed forever, and I mean this literally! Once exposed to other channels and frequency realms, our basic physiology is changed forever. Once you have been wired to perceive a new "channel", you, to some degree, cannot stop perceiving it!

I want to say more about this final conclusion. In every one of the cases presented in this book, my witness' lives, their relationships, and their understanding of their world was dramatically changed. Many of my witnesses whom I have studied over the years seem to fall into two categories:

They have been sensitive from birth (i.e., they have already been aware of more "channels" than the people surrounding them since birth, so they live in worlds that actually contain more information than their peers) or they have a defining moment in their lives where they are ushered into these other worlds of perception (i.e., in my world of study, they have a dramatic UFO/alien event that fundamentally rewrites their very being and actually introduces them, sometimes forcibly, into other channel awareness).

Spoiler Alert: In a future book, I will explore the relationship between these categories of knowing and Sympathetic Resonance versus Entrainment. This book will explore the ways we can actually learn about precision of state change and channel reception, meaning, we can learn to tune into these other worlds and beings!

Let's really think about these two choices of indoctrination into other realms. When you come into the world more aware than your parents and family, you have a very high probability of being dismissed, ridiculed, and alienated. The people around you are all living on channels 1-3, and you live effortlessly on channels 1-6. You see and possibly interact with UFOs and their occupants.

Sometimes you experience ghostly phenomena, strange cryptids/beings, and are extremely sensitive to others' emotions and moods. No one will listen to you, and, worst of all, you are ridiculed, dismissed, and possibly punished for your abilities. This happens also to the people who come into their abilities later in life as described in category 2.

I had one witness who, when she was a little girl, experienced dishes flying off the shelves in the kitchen when she got upset or scared. One day when this happened, her mother was angry and yelled at her to "stop showing off". That statement alone intimated at the fact the mother knew what was going on to some degree. Due to her own fear, either of society's or the family's reaction, the mother shamed the child into hiding her abilities. So, guess what the child did? She stopped, as all of us will give up our true selves for love. And, with this cessation of her true self, all of her cool abilities and perceptions were buried under a sea of repression. Also to be noted, with the huge burden of hiding her true self away, of repressing large amounts of energy and information, that energy stagnated, causing lifelong emotional and physical problems. This pattern is repeated ad nauseam throughout my case files.

Think about this, really think! The very people in our society who have advanced evolutionary abilities have been and continue to be denigrated and ostracized. It's really insane how so many of my kind, sane witnesses are treated. We live in a society where our judicial system is based on credible witness testimony, is taken very seriously, and will either convict or acquit someone of a crime. But take that same individual and have him report a UFO sighting or, heaven forbid, some paranormal event, and he will be made fun of or outright not believed. I started one of my lectures the other day with this, "I'm going to make up a story today that I saw a UFO and furthermore I saw an alien also. I'm now going to tell my family, friends, and work colleagues so they can all make fun of me for the rest of my life." Makes no sense at all that people who have been known their whole lives as steady, rational individuals would just, on a whim, create an outrageous story so they can be maligned forever about it. The very individuals we should be studying and listening to are marginalized and labeled as crazy, and we just keep crashing onward with the blind leading the blind.

Thank goodness, many years ago something grabbed me by the scruff of the neck and forced me to pay attention. I continued to study in the known realms, but when something did not fit into that comfortable world view, and it kept showing up over and over, I listened. And, thank goodness so many other grassroots researchers paid attention, too. Recently, many fellow serious investigators from different paranormal fields have started coming together to share their work as they have clearly discovered the crossover events that happen in their respective fields. For example, paranormal researchers who mainly focus on ghostly phenomena will have UFO sightings, UFO researchers will have witnesses report UFO/cryptid encounters, and the cryptid researchers will began to report strange lights and paranormal activity. The serious researchers have known for a long time that there is a connection between all of these mysterious phenomena. What has become apparent after years of research focusing solely on the external events themselves is that in many ways, most investigators are no closer to the truth and, in reality, have even more questions. In changing the way I researched early on by including the study of my witnesses in conjunction with their sightings/events, I discovered what I feel is the fundamental connection between all of these mysteries. We as humans already possess some incredible abilities, and with awareness and training, can push the development of those abilities into sets of refined skills. Why we as a species overlook the continued evolution of human consciousness is beyond me. Over 35 years of research with some astonishing individuals has shown me otherwise. *Our consciousness and the advanced abilities of many individuals to perceive along a continuum of the electromagnetic spectrum whether acquired from birth, trauma, or trained, IS the connection between all of these phenomena.*

With recent government admission of the study of UFOs, NASA's quiet releases of new exoplanets that are similar to Earth and an enormous entertainment industry cranking out paranormal/ET shows, I feel we are on the cusp of some major paradigm shifts for society. I hope as we go forward, more people can be open to the incredible possibilities that lay ahead of humanity. That we as a species can lay aside our fear of the unknown long enough to learn something new. That we can go forth with a spirit of curiosity, actually listen to our friends and loved ones when they report something strange, and stay open long enough to honestly consider their experiences. I am forever grateful for the courage, honesty, and resilience of all of my witnesses. They are my family and friends. I am humbled, amazed, and thrilled about the worlds they have seen, the abilities they now possess, and the promise of transformation they all exemplify. They have truly gone into that "Far Country", have risked family, friends, and careers to speak out and share with us their tales of radical new worlds. I invite my readers to exceed their fears and prejudices, to strike out on their own "Great Journeys", and to join us in the exploration of what I feel is the next great frontier.

- *UFOgirl…Tennessee*

REFERENCES

Clelland, Mike, *The Messengers: Owls, Synchronicity and the UFO Abductee*, Richard Dolan Press; 1st edition (December 3, 2015).

Noel, Christopher, *Mindspeak, Tapping into Sasquatch and Science.* 2019, by Christopher Noel, pp.110.

Gordon, Stan, *Silent Invasion: The Pennsylvania UFO-Bigfoot Casebook, 2010*

The Monroe Institute. For more information, see: https://www.monroeinstitute.org/

Monroe, Robert, *Journeys out of the Body*, (Doubleday, 1971)

Monroe, Robert, *Far Journeys*, (Doubleday, 1985)

Monroe, Robert, *Ultimate Journey*, (Doubleday, 1994)

Wilbur, Ken, *Integral Spirituality*, (Shambhala Publications, Inc., 2006)

Howe, Linda Moulton, *A Strange Harvest,*

Meldrum, Jeff, *Sasquatch: Legend Meets Science,* Sept. 2007, Tom Doherty Assoc.

Paulides, David, *The Hoopa Project*, 2008, Hancock House Publishers LTD.

Walton, Travis, *"Fire in the Sky"*, (Marlowe & Company, 1979, 1996)

Dunn, B.A. C.H.T., Tonya: Hypnosis Session, Witness "J"

Bakara, David, Expedition Bigfoot Museum, 1934 GA-515, Cherry Log, GA 30522

Delph, Matthew, *Founder of: MECRO, Mountain Empire Cryptid Research Organization*

From "Tennessee History" Tennessee Good Old Days https://tennesseehistoryblog.wordpress.com/2017/08/03/palmersvi lle-and-latham-tn-wwii-b-17-crash-sept-1943/comment-page-1/

ABOUT THE AUTHOR

Angelia Sheer has been investigating the UFO mystery for 35+ years. She founded Parasheer Research, her private research group, and is also the current State Director/Chief Field Investigator and STAR Team Lead for MUFON of Tennessee. With her extensive years of "boots on the ground" field work, over 2,500 witness interviews, and her dogged pursuit of the truth, Angelia was named MUFON's 2019 Field Investigator of the Year.

Angelia is also a certified master hypnotherapist and uses her training to aid experiencers in dealing with their UFO encounters. She has made it her priority to provide a safe, confidential place for witnesses to share their stories and to discover other individuals who have had similar sightings and experiences. She shares her incredible cases via live speaking engagements, radio shows, TV episodes, websites, and various other media. It has been an honor for her to have so many witnesses come forward with trust and reveal their exciting, mystifying, and sometimes frightening adventures into the unknown.

When Angelia is not out chasing UFOs, she spends her time hanging out at her farm in Tennessee where she has been training and riding horses for over 35 years.

For more information, visit http://www.angeliasheer.com

www.ingramcontent.com/pod-product-compliance
Lightning Source LLC
Chambersburg PA
CBHW021354210526
45463CB00001B/98